BOTANY
A JUNIOR BOOK FOR SCHOOLS

BOTANY

A JUNIOR BOOK FOR SCHOOLS

BY

R. H. YAPP

CAMBRIDGE
AT THE UNIVERSITY PRESS
1953

CAMBRIDGE UNIVERSITY PRESS
Cambridge, New York, Melbourne, Madrid, Cape Town,
Singapore, São Paulo, Delhi, Mexico City

Cambridge University Press
The Edinburgh Building, Cambridge CB2 8RU, UK

Published in the United States of America by Cambridge University Press, New York

www.cambridge.org
Information on this title: www.cambridge.org/9781107619548

First edition 1923
Second edition 1924
Reprinted 1925
Third Edition 1927
First published 1927
Reprinted 1929, 1930, 1938, 1943, 1946, 1949, 1953
First paperback edition 2013

A catalogue record for this publication is available from the British Library

ISBN 978-1-107-61954-8 Paperback

PREFACE

IT has been said, and with more than an element of truth, that the least important part of education is the acquisition of knowledge. The facts of nature, as such, have an intense fascination for the nature-lover, yet the value of a mere knowledge of facts may easily be over-estimated. Facts are really tools, the value of which lies in the uses to which they may be put. Of far deeper importance, therefore, than the mere committing of facts to memory, is the acquisition of correct habits of study. Habits, that is, of patient and accurate observation, and of clear and logical interpretation and correlation of the facts observed. Facts may be forgotten, but habits remain.

In the hope that I might be able, in some measure, to second the efforts of teachers to help their pupils to acquire such habits, I have endeavoured to keep three chief aims before me in writing this book: (1) By employing as far as possible the inductive method, to make facts lead up to and illustrate principles. (2) To maintain a logical sequence in the arrangement of the subject-matter. This is why climbing plants are included in the chapter on "How Foliage Leaves get Light," and why the chapters dealing respectively with the movements of plants and with water plants follow that on respiration. (3) To make accuracy and thoroughness the key-note of the book, so far as the limits of space allowed.

The drawings (all of which are original), as well as the descriptions, have been made, as I hope those of pupils using this book will always be made, from actual specimens. Care has been taken to make the drawings as accurate as possible. In most cases the specimen has been drawn in the position in which it occurred in nature (*e.g.* Figs. 99, 100 A, 120). Where underground parts are shown, the soil level has usually been indicated—this is often important (cf. Figs. 101–103). The month in which the drawing was made, and the magnification used, are recorded between brackets underneath most figures. The original drawings are on a much larger scale than that on which they could be reproduced in this book. It is advisable for the pupil to cultivate the habit of making drawings, especially those of small objects such as flowers, on a really large scale. This renders it far easier to record correctly not only the shape, but also the proportions, of the various parts, and the relations of these parts one to another.

As regards subject-matter, it need only be said that the aim has been to select matter suitable for beginners, and to avoid unnecessary overlap with courses taken at a later stage. The physiological experiments described are therefore for the most part simple, and qualitative rather than quantitative. The "out-of-doors" aspect of plant life has been emphasized, but all microscopic structure omitted. Chapters XVIII to XXI will, I hope, arouse interest in the sometimes neglected study of seasonal changes in plants. The book is intended to provide a sound course of instruction in the fundamental principles of Botany. It should be found amply sufficient for those preparing for the Preliminary and Junior Local Examinations of the University of Cambridge, or for similar examinations. For matriculation purposes it will probably need to be supplemented in certain directions.

It only remains gratefully to tender my thanks to friends who have helped me during the writing of this book. In the first place the book owes much to the many suggestions and sound criticisms of two of my former students, Miss M. M. Wells and Miss E. J. Rutledge, whose good judgment, and experience of the problems of actual school teaching, have enabled me to avoid many pitfalls into which I might otherwise have fallen. To Professor A. C. Seward, who long ago suggested the writing of this book, Dr Harold Wager, H.M.I., and Miss M. Scott I am also indebted for criticisms ; also to Mr A. G. Tansley for permission to reproduce Fig. 70, from *The New Phytologist*, and to Dr Klintberg for kind assistance in proof-reading. And last but not least I wish to acknowledge the great help afforded by my wife, who has criticised my style, typed my MS., compiled the index, and given that encouragement without which, in the press of other duties, the book might never have been completed.

R. H. Y.

BOTANICAL DEPARTMENT,
UNIVERSITY OF BIRMINGHAM.
July, 1923.

PREFACE TO THE SECOND EDITION

ORIGINALLY it had been intended to include the Latin as well as the English names of the chief plants mentioned, but the Latin names were finally omitted as perhaps unnecessary in an elementary book. Since publication, however, it has been suggested that the inclusion of at least some Latin names might be desirable. There is no doubt that familiarity with the scientific names of plants from an early stage is a help to those who may ultimately become more deeply interested in the study of Botany. An appendix has therefore been added, giving a number of English plant-names with their scientific equivalents, and a few of the more obvious derivations. The other changes in this edition consist of a number of minor corrections and additions in the text and index, and the introduction in Chapter XXVIII of a paragraph on the use of a "Flora."

I am again indebted to several of my friends, particularly to Professor Seward, for useful suggestions and criticisms; also to Professor J. O. Thomson for his kind help in connexion with the Latin derivations.

R. H. Y.

September, 1924.

PREFACE TO THE THIRD EDITION

IN this edition a few modifications and additions have been introduced, several of them as the result of suggestions by Miss M. M. Wells, to whom my thanks are due. The changes, however, are not of such a nature as to interfere with the use of this and previous editions side by side in the same school.

R. H. Y.

May, 1927.

CONTENTS

CHAPTER 1

THE GROUNDSEL, THE SYCAMORE AND THE COCK'S-FOOT GRASS

Plants of very many kinds are found all over the world. Some live on land, others in the water. Some are only to be found in woods, while others grow on the seashore, or even in dry, stony deserts. There are indeed very few places on the surface of the globe where plants of some kind or other are not to be found. Let us take a few of these plants and find out as much as we can about them.

The Groundsel, a very familiar weed, will do to start with. It is to be found in almost every garden, but you will look for it in vain in such places as a grassy meadow, an Oak wood, or in the water.

Look at a Groundsel plant as it grows in a garden. The green part above the soil is called the **shoot**; it consists of an upright, rather juicy **stem,** and a number of flat **leaves** (Fig. 1). Towards the upper end of the stem you may find some yellowish **flowers,** or even a cluster of small **fruits,** each with a tuft of hairs. If you dig up the plant and shake it free from soil, the whitish underground parts, or **roots,** will be seen.

Now examine the plant a little more carefully.

1. **The shoot.** The **main stem** is rounded, especially at its lower end. Higher up it is ridged, and has an angular appearance. The parts where the leaves are joined to the stem are called **nodes,** the smooth parts between the nodes being the **internodes.** In the Groundsel there is only one leaf at each node, and the leaves are arranged round the stem in a spiral. If we look at a plant first from the side and then from above (Figs. 1, 2), we see that this arrangement makes it easy for every leaf to get light and air. Later on we shall learn that this is of the greatest importance.

In the lower leaves we can distinguish three parts: (*a*) the flat, thin **leaf-blade,** the edge of which is irregularly toothed;

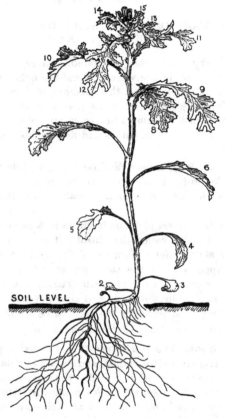

Fig. 1. GROUNDSEL PLANT. Leaves numbered according to age, the oldest below, the youngest above. (August, × 1/2.)

(*b*) the narrow **leaf-stalk** or **petiole,** and (*c*) the slightly broader **leaf-base,** by which the leaf is joined to the stem. The leaves of the Groundsel vary both in size and shape, even on the same plant (Fig. 3). The upper leaves, unlike the lower, usually have

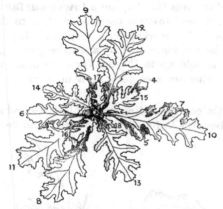

Fig. 2. Groundsel Shoot seen from above. (× 1/2.)

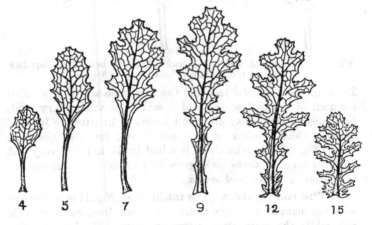

4 5 7 9 12 15

Fig. 3. Groundsel: Different Shapes of Leaves on the same Plant. Numbered as in Fig. 1. (× 3/4.)

no petioles; their blades are also more deeply toothed or lobed, and often clasp the stem (Fig. 4). Each leaf has an **upper surface** turned towards the sky, and a **lower surface,** of a lighter green colour, turned towards the ground. The **midrib** or **main vein** of the leaf is marked by a projecting ridge on the lower surface of the leaf-blade, and by a shallow groove on the upper. If the leaf is held up to the light and examined with a pocket lens, other veins are seen forming a network in the leaf-blade (Fig. 3).

Buds. On gently bending down a leaf, we find a small, hairy **bud** just above the leaf-base, and between it and the stem (Fig. 4)

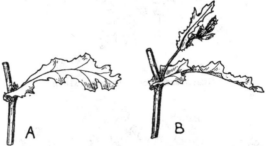

Fig. 4. GROUNDSEL. A, a node with leaf and axillary bud. B, axillary bud grown into a branch shoot. (Aug. × 3/4.)

The angle between the leaf and the internode above it is called the **axil** of the leaf, so the bud is said to be an **axillary bud.** Later on this bud may grow and become a **branch** or **lateral shoot,** which consists of stem and leaves, just like the main shoot (Fig. 4). Usually there is a bud in the axil of every leaf, but only some of these buds grow into branches. Branches can only arise in the axils of leaves.

2. **The root-system.** The **main root** (Fig. 1) is continuous with the main stem of the plant. It grows downwards into the soil, while the main stem grows upwards away from it. The **secondary** or **lateral roots** are branches of the main root.

They spread outwards from the main root, growing between the particles of soil. The secondary roots give rise to still smaller **rootlets** (Fig. 1). The roots of the Groundsel branch a great deal, but they only form other roots, and never produce leaves, buds or flowers. By means of its roots the plant is anchored to the spot where it is going to spend the rest of its life.

The Groundsel plant then consists of a number of different parts or **organs.** We shall see later that each of these organs has some special work to do. The plant usually grows quickly, and dies after producing flowers and fruits. Several generations of plants may be produced in a single season, so that we can find the Groundsel at almost any time of the year. The Groundsel does not live for more than one year, and therefore is said to be an **annual.**

The Sycamore Tree. We may next examine the Sycamore tree. At first sight such a big plant (Fig. 5) seems altogether different from the humble Groundsel, but even the tallest tree was small once, so if we start with a young Sycamore we can more easily compare it with the Groundsel. The plant shown in Fig. 6 is about a year and a half old. We see at once that there is a branching root-system, firmly fixed in the soil. Like those of the Groundsel, the roots of the Sycamore produce neither leaves nor buds, nor are they green in colour. Above the ground is the shoot, with an upright stem, stronger than that of the Groundsel, but divided into nodes and internodes in the same way. Notice that in the Sycamore there are two leaves instead of one at each node. The leaves are different in shape from those of the Groundsel, but resemble them in having flat, green leaf-blades and axillary buds. The petiole is very distinct from the leaf-blade and is joined to the stem by the leaf-base.

The midrib and other veins branch again and again, forming a beautiful and delicate network in every part of the leaf-blade. Many of the smallest branches end as little free tips within the meshes of the network (Fig. 7).

The Groundsel is an annual, but the Sycamore lives for a great many years, and is therefore called **a perennial.** Each year the roots increase in size, and become more and more numerous. The main stem, too, becomes very tall and thick, forming the woody **trunk** of the tree (Fig. 5). Every year some of the buds grow into new branches, so that an old tree may have hundreds of

Fig. 5. SYCAMORE IN WINTER CONDITION. (March.)

branches of all sizes. The smallest branches are known as twigs. All these branches, even the largest ones, began life as little buds in the axils of leaves. In the Sycamore, as well as in the Groundsel, branches only arise in the axils of leaves.

Appearance at different seasons of the year. During the summer every small twig carries several pairs of leaves, so that the whole tree is covered with green foliage. In the autumn,

however, the leaves become yellow and finally fall off, leaving
smooth **leaf-scars** on the twigs. In the axils of these leaf-scars
we can still see the buds, which remain behind when the leaves
fall (Fig. 96).   At the tip of each twig is a **terminal bud**,
which did not arise in the axil of a leaf. Thus buds may be
either terminal or axillary in position. All the buds are covered

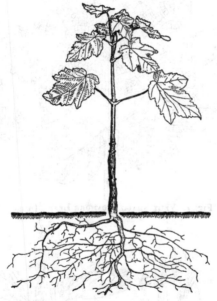

Fig. 6. YOUNG SYCAMORE. (Aug. × 1/4.)

with small, overlapping **bud-scales** (Fig. 88). The tree does
not grow during the winter-time, but remains in the bare, resting
condition which is so familiar to us (Fig. 5). In the early spring,
however, about March or April, growth begins again.   The
numerous terminal buds, and some of the axillary ones, swell,
becoming longer and longer, as the bud-scales and the delicate

Fig. 7. VERY SMALL SYCAMORE LEAF. (Aug. × 2.)

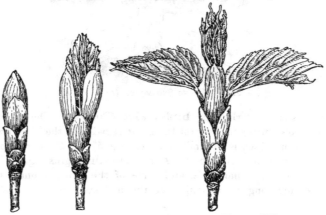

Fig. 8. OPENING BUDS OF SYCAMORE. (May, × 5/8.)

parts inside begin to grow. Soon the buds burst, and the small
folded leaves appear (Fig. 8). In a surprisingly short time the
whole tree is once more covered with leaves. One reason why
this happens so quickly is that the leaves were really formed
during the previous summer. They have rested during the winter
inside the buds, and when spring comes all they have to do is to
burst the buds and grow to their full size.

The Cock's-Foot Grass may be taken as a third example.
This is a rather large grass, commonly found in pastures and
waste places and on railway banks, where it often forms coarse,
rough tussocks. If we look at one of these tussocks, we find it is
made up of a great many green shoots. Unlike the Groundsel
and the Sycamore, however, this plant does not appear to have
any main stem. Also, if we dig up the plant, we find many
branching roots in the soil, but no main root.

Let us carefully examine one or two of the shoots which make
up a tussock (Fig. 9). Each shoot has a stem which, like those
of the other two plants we have studied, is made up of nodes and
internodes. One leaf is inserted at each node. The lower part of
each stem usually takes up a creeping position, but the upper
part is erect. In the lower part the leaves are crowded together,
and the internodes are short, while the erect part bears fewer
leaves, and the internodes are longer. In the summer-time flowers
are formed at the top of the stem (Fig. 9).

The leaves are quite different in appearance from those of both
the Groundsel and the Sycamore. Each leaf has a long, narrow
leaf-blade, and a long leaf-base which forms a sheath right round
the stem or the younger leaves (Figs. 9, 10). There is no petiole.
Just where the leaf-blade joins the sheathing leaf-base, you will
notice a curious papery structure which lies flat against the stem,
or else against the sheath of the leaf above. This is called the
ligule (Fig. 10). Its duty is to prevent rain from running down
inside the sheath. Each leaf has a number of veins, which are
parallel to each other (Fig. 10), and do not form a network as do

those of the Groundsel and Sycamore. If you open a leaf sheath and carefully pull it down as far as it will go, you will find that

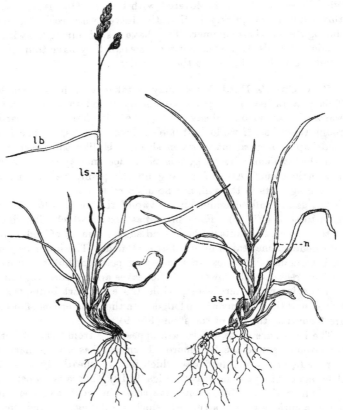

Fig. 9. Cock's-Foot Grass. *as*, axillary shoot; *n*, node; *lb*, leaf-blade; *ls*, sheathing leaf-base. (Sept. × 1/5.)

the node, where the leaf is actually joined to the stem, is some distance below the ligule. Notice the small, pointed bud in the leaf-axil, hidden away inside the sheath (Fig. 11). Later on, some of these axillary buds will develop into new branch shoots (Fig. 9 *as*). The Cock's-foot Grass is a perennial, and forms new axillary branch shoots each year. In this way big tussocks are gradually built up. Unlike the Sycamore, the Cock's-foot, and most other grasses, have green leaves all the year round.

The roots of the Cock's-foot Grass are more wiry than those of the Groundsel. You will notice that they grow out of the creeping

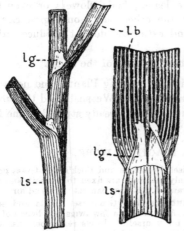

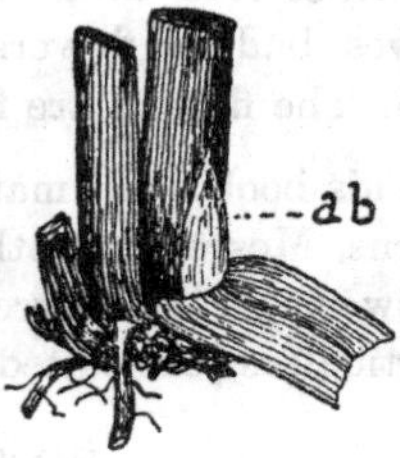

Fig. 11. Cock's-Foot Grass. Axillary bud (*ab*) at base of stem. (Aug. × 1.)

Fig. 10. Cock's-Foot Grass. *lb*, leaf-blade; *ls*, sheathing leaf-base; *lg*, ligule. (Aug. × 1½; right Fig. × 3.)

part of the stem, and not from a main root. Roots which grow from stems in this way, and not from other roots, are called **adventitious roots.** Adventitious roots are found in many plants; e.g. in the Groundsel, in addition to the ordinary roots, we often find adventitious roots growing amongst the leaves on the lower part of the stem (Fig. 1).

Summary. The Groundsel, Sycamore and Cock's-foot Grass differ from one another in size, length of life, shape of leaves and many other respects. But they are all built on the same general plan, and resemble each other in the following points :

1. There is a green shoot-system above the ground, and a white or brownish root-system buried in the soil.

2. The shoot consists of a stem bearing leaves at its nodes. A leaf is always borne on the side of a stem, and usually has a bud in its axil.

3. Branches or lateral shoots only arise from axillary buds.

4. The stem may produce leaves, buds, flowers or even adventitious roots. Roots, on the other hand, only give rise to branches like themselves, and as a rule do not produce either leaves, buds or flowers.

5. The flowers are found near the tips of the stems.

This book deals mainly with "Flowering Plants" and not with Ferns, Mosses and other lower plants. We shall find that most Flowering Plants agree with the three already studied in the five particulars just stated.

HINTS FOR PRACTICAL WORK.

1. Look for plants of Groundsel, Sycamore and Cock's-foot Grass growing out of doors. Notice how firmly the roots are fixed to the soil, and how in each case the leaves are so arranged that they can get light and air.

2. Dig up carefully one or two complete Groundsel plants and some shoots of the Cock's-foot Grass. Also collect a few twigs and leaves of the Sycamore. You should do this for yourself. Always remember that *it is very important to examine actual specimens while you are reading,* whenever this is possible. You will no doubt be able to find out other things about the plants for yourself besides those which are described in this book. Notice, for instance, whether the adventitious roots of the Cock's-foot Grass grow from the nodes or from the internodes of the creeping part of the stem.

3. *When you are quite sure you understand a specimen, make careful drawings of the things you see.* These should be made from the specimens themselves and not from the drawings in the book.

4. Examine a few other plants, such as an Oak tree, Wallflower, Broad Bean and Chickweed, and notice that the five points given in the summary are true of these plants too.

CHAPTER II

FLOWERS AND FRUITS

Leaves, stems and roots are called the **vegetative organs** of a plant. They are, as we shall learn later, necessary for the growth and well-being of the plant itself. Flowers, on the other hand, contain the **reproductive organs,** that is, the parts which have to do with the production of new plants. Let us examine one or two flowers.

Fig. 12. INFLORESCENCE OF DAME'S VIOLET. (July, × 3/4.)

The Dame's Violet. We may begin with the flowers of the Dame's Violet, which is often grown in gardens. If this is not available, the flowers of the single Stock, Wallflower, or Lady's Smock, which are very similar, will do just as well. Except for colour and time of flowering, the following description will do for any of these flowers.

The Dame's Violet blooms from about May to July, the flowers being clustered together towards the tips of the branches. Such a cluster of flowers is called an **inflorescence** (Fig. 12). Each flower has a short **flower-stalk,** at the top of which are the different parts which make up the flower itself. These parts are really **floral leaves,** though they are very different from the green **foliage leaves** considered in the last chapter.

On the outside of the flower are four small green leaves

called **sepals** (Fig. 12), two of which are slightly enlarged at the base to form little pockets. Just inside the sepals are four larger leaves, the **petals,** of white or lilac colour arranged like a cross. Pull off a petal, and notice its shape. There is a narrow part hidden by the sepals and a broad, flat part above them.

Next come six **stamens,** which are best seen if you pull off all the sepals and petals. Each stamen has a stalk called the **filament,** and a swollen tip known as the **anther** (Fig. 13). If you examine a ripe anther with a pocket lens, you may see that it has split open along two lines, and that a fine powder

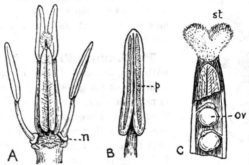

Fig. 13. DAME'S VIOLET. A, flower with sepals, petals and two stamens removed; *n*, nectary. B, an anther; *p*, pollen. C, part of ovary cut open to show two ovules (*ov*); *st*, stigma. (July, A × 4; B, C × 8.)

(called **pollen**) is coming out through the openings (Fig. 13 B). Finally, in the very centre of the flower is the **ovary,** a stalk-like structure ending in two sticky **stigmas** (Fig. 13, A and c). Split open a rather old ovary, and notice a number of very minute roundish **ovules** inside, joined to the ovary wall by short stalks (Fig. 13 c). Each ovule is really an unripe seed.

Take another flower and notice that two of the stamens are shorter than the other four. At the base of the filament of each short stamen is a little green swelling, the **nectary** or **honey-gland.** The honey made in the nectary is stored in the pockets of the two sepals which are opposite to the short stamens. In sunny weather you may often see bees or other insects visiting the flower to get this honey.

Now look at the older flowers at the bottom of the inflorescence. From some of these the sepals, petals and stamens may have fallen, but the ovary is still there. Gradually the ovary grows and becomes a hard, dry **fruit,** many times longer than it was at first. When quite ripe the dry fruit splits into three parts and the small, oblong **seeds** escape (Fig. 14).

The real work of the flower is to form seeds, each of which may grow into a new Dame's Violet plant. This is how the plant reproduces or multiplies itself, so that when the parent plant dies, there will still be young ones left. Later on we shall learn how the various parts of the flower—sepals, petals, stamens, pollen, ovary and stigmas, as well as the honey and the insects which come for it, all help in the production of seeds.

The Broad Bean. The flower of the Broad Bean is very unlike that of the Dame's Violet, yet it has the same kinds of organs. There is a short flower-stalk, at the top of which are five sepals. Except at their tips, the sepals are all joined together, forming a kind of cup (Fig. 15). Just inside the sepals are five petals. In the Dame's Violet the four petals are all alike, but in the Broad Bean we find three different kinds of petals. These are (*a*) one large **standard** petal at the top of the

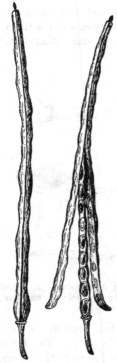

Fig. 14. RIPE FRUITS OF DAME'S VIOLET. (Aug. × l.)

flower, (*b*) two side petals or **wings,** and (*c*) two lower petals joined together along their edges to form what is called the **keel** (Fig. 15). The petals are white, except for some brown lines on the standard, and a very large brownish-black spot on each of the wing petals.

The stamens and ovary are hidden away inside the keel. If you pull off the sepals and petals very gently you will see them. There are ten stamens, each with its filament and anther. The

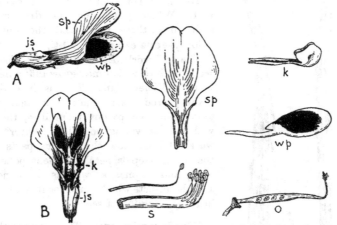

Fig. 15. BROAD BEAN. A, flower. B, flower seen from below. *js*, joined sepals; *sp*, standard petal; *wp*, wing petal; *k*, keel; *s*, stamens; *o*, ovary with style and stigma. (Aug. ×1; *s*, *o*×1½.)

lower parts of nine of the filaments are joined together to form a little trough (Fig. 15 *s*), in which honey is stored. The tenth stamen (the one opposite the standard) is not joined to the others, but forms a kind of loose lid to the trough. Bees are able to get the honey by inserting their tongues between the lid and the trough.

The ovary is a slender structure lying in the trough formed by the stamens. It is nearly transparent, so if you hold it up to the light, you can count the ovules inside. The ovary ends in a bent portion, the **style,** at the top of which is a minute stigma. The Dame's Violet also has a style between the stigmas and the ovary, but it is very short, and not nearly so distinct as in the Broad Bean, though the stigmas are more distinct (Fig. 13).

After a time, the sepals wither and the petals and stamens fall off. The ovary, however, grows and develops into the large bean-pod or fruit (Fig. 16). When quite dry and ripe, the fruit splits into two halves and allows the seeds to escape.

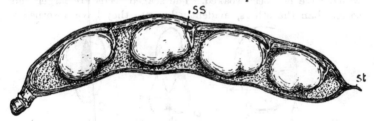

Fig. 16. Fruit of Broad Bean. One side of ovary cut away to show the seeds. *ss*, seed-stalk; *st*, stigma. (Aug. × 1/2.)

HINTS FOR PRACTICAL WORK.

1. Examine flowers and fruits of the Dame's Violet and Broad Bean, and draw the various parts. If you cannot get these flowers, you can use one of those mentioned above instead of the Dame's Violet. In the same way, the Sweet Pea, Garden Pea, Runner Bean, Lupin, Laburnum or Vetch, may be used instead of the Broad Bean.

Compare the flowers of the Dame's Violet and the Broad Bean, making two lists of characters, i.e. (*a*) those in which the two flowers resemble, and (*b*) those in which they differ from each other.

2. You will find a sharp pocket knife and two strong needles very useful in dissecting flowers or other parts of plants.

3. *Small parts should always be examined with a pocket lens.* You will see many structures more clearly with a lens than with the naked eye.

4. Examine a few other flowers and fruits, e.g. Buttercup, Primrose and Poppy. Notice that in each case (*a*) the flower has sepals, petals, stamens, and one or more ovaries; (*b*) after the flower withers, the ovary develops into the fruit, and (*c*) until the seeds are ripe, they are sheltered inside the fruit.

CHAPTER III

SEEDS

Seed of the Broad Bean. Soak a few Broad Bean seeds in water for a day or two, and then compare them with seeds which have not been soaked. The soaked seeds are larger and softer than the others, evidently because they have absorbed a

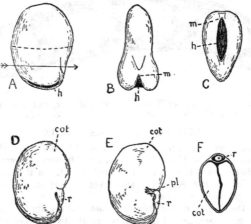

Fig. 17. SEEDS OF BROAD BEAN. A—C, seen from outside; *h*, hilum; *m*, micropyle. D, embryo after removal of testa. E, embryo with one cotyledon removed. F, seed cut along arrow in A. *cot*, cotyledon; *r*, radicle; *pl*, plumule. (A, B, D, E × 3/4; C, F × 1.)

good deal of water. At the thicker end of the seed is a long, dark-brown scar (the **hilum**), showing where the seed-stalk, which joined the seed to the pod, has broken away (Figs. 17, 16). At one end of the hilum is a minute slit-like hole, called the

micropyle (Gk. *micros*, small, *pyle*, a gate). If you dry the outside of a soaked bean, and then gently squeeze it, water comes out through the micropyle, showing that there really is a hole there.

The seed is covered by a tough skin, the seed-coat or **testa.** This can easily be removed from one of the soaked seeds by first cutting through the testa with a sharp knife in the direction of the dotted line (Fig. 17 A), and then peeling it off in two halves. Underneath the testa is a fleshy, cream coloured mass, which fills the rest of the seed. This is the young plant or **embryo,** as it is called. The embryo has two thick seed-leaves or **cotyledons,** lying against the outside of which is a small, pointed root or **radicle** (Fig. 17 D). Notice that the radicle fits into a little pocket on the inside of the testa (Fig. 17 F); the micropyle is opposite the bottom of this pocket. On separating the two fleshy cotyledons, you will see a small bud lying between them; this is the young shoot or **plumule.** Examine it with a lens, and notice that it consists of a minute stem bearing folded leaves (Fig. 17 E).

Cut one of the cotyledons open and put a little solution of iodine on the cut surface; a bluish-black colour is produced. If you add iodine to the cut surface of a Potato, or to a piece of washing starch, the same bluish-black colour is seen. Iodine always gives this colour when it comes into contact with starch, so it forms a useful test for the latter substance. Starch, which is an important food for both plants and animals, is found not only in the Bean and Potato, but in many other plants as well.

We have now found out that:

1. The Broad Bean seed consists of an embryo covered by a testa.

2. The embryo, like the older plants we have studied, is made up of stem, leaves and root.

3. The two large fleshy leaves (cotyledons) of the embryo contain a store of starchy food.

Seeds of the Sycamore and Ash. In the autumn when
the leaves of the Sycamore and Ash are falling, many of the trees
are seen to be nearly covered with bunches of brown, winged
fruits, or "keys" as they are often called. Gradually these fruits
are scattered by storms and wind, the Sycamore fruits usually
falling earlier than those of the Ash.

Each **Sycamore** fruit has two (sometimes three or more)
thin flat wings (Fig. 18), at the base of each of which is a swollen

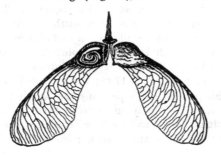

Fig. 18. Winged Fruit of Sycamore. Ovary
wall on left is cut away to show coiled embryo.
(March, × 1.)

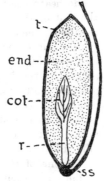

Fig. 19. Seed of Ash,
cut open to show
Embryo. *t,* testa;
end, endosperm; *cot,*
cotyledon; *r,* radicle;
ss, stalk of seed.
(Nov. × 3.)

part, lined with hairs and containing a
single seed. The seed is covered by a
thin brown testa. If you scrape away
the testa you will find the embryo inside,
consisting, as in the Bean, of a radicle,
two cotyledons and a very minute plumule. The cotyledons are
green and long and narrow, and are coiled into a sort of ball
(Figs. 18, 22); they too contain a store of food.

The **Ash** fruit has only a single wing, and contains one flat
seed (see Fig. 145 A–C). The testa is thin and brown, and cannot
easily be peeled off. If we carefully slice away part of one of the
flat surfaces of the seed, we find a small white embryo inside
(Fig. 19). In this case also, the embryo consists of a radicle, two

cotyledons and a plumule, though, on account of their small size, it is often rather difficult to make out some of the parts until later, when the seed germinates.

In the Bean and Sycamore seeds, the embryo fills the whole of the space inside the testa. In the Ash, however, the embryo occupies only part of this space, the rest being filled with a semi-transparent, horny store of food, known as **endosperm** (Fig. 19).

Dicotyledons and Monocotyledons. Flowering Plants may be divided into two great groups, in one of which the embryo has two cotyledons, and in the other only one. On this account, the groups are known as Dicotyledons and Monocotyledons respectively. The Bean, Sycamore, Ash, and indeed most of our common plants, are Dicotyledons, while grasses, Daffodils, Tulips etc. belong to the Monocotyledons. We will now take an example of a Monocotyledon.

Maize grain. The grain of the Maize is usually flat and yellow, and is covered by a thin skin. On one of the flat sides

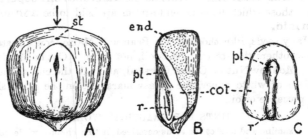

Fig. 20. MAIZE. A, whole grain; *st*, remains of stigma. B, grain cut vertically. C, embryo dissected out. *cot*, cotyledon; *pl*, plumule; *r*, radicle; *end*, endosperm. (× 2¼.)

may be seen an oval whitish patch, showing where the embryo lies just below the skin (Fig. 20 A). If you cut a grain exactly through the middle, in the direction of the arrow in A, Fig. 20, you can see the embryo, with its plumule, radicle, and single cotyledon (B). The rest of the grain is filled with endosperm, which, on testing with iodine, is found to contain starch. In order to understand the shape of the embryo better, soak a few grains of Maize in water for a day, and then keep them on damp blotting paper in a covered dish for four or five days longer. You can now easily separate the embryo from the rest of the grain. Note the large folded cotyledon, which wraps almost completely round the radicle and plumule (Fig. 20 c).

The seeds we have studied differ from each other in shape, size and colour, but they, and indeed all seeds, agree in three respects :

1. They all contain an embryo or young plant.

2. They all contain a store of food.

3. In each case there is a seed-coat or testa[1].

The stored food. We have seen that in the Ash and Maize food is stored in the endosperm, i.e. outside the embryo, while in the seeds of the Broad Bean and Sycamore the food is stored in the embryo itself (in the cotyledons), and there is no endosperm. Seeds which have endosperm are called **endospermic,** while those which have no endosperm are said to be **non-endospermic.**

Many seeds, like those of the Bean and Maize, contain starch, but in others the stored food is oil, not starch. Cut open a Brazil Nut, Monkey Nut or Castor Oil seed, and rub the cut surface on a piece of white paper, the greasy mark left on the paper shows the presence of oil.

HINTS FOR PRACTICAL WORK.

1. Examine and dissect the seeds described in this chapter while you are reading it. Make drawings from your own specimens. Examine other seeds as well, and find out all you can about them. The Pea, Runner Bean, Butter Bean, Castor Oil, Buckwheat, Acorn, Barley, Hazel Nut, and Sunflower are all good ones to use.

2. Write a comparison of the structure of all the seeds you have examined, pointing out clearly their resemblances and differences. In each case state where the food is stored and whether starch or oil is present.

[1] It is not easy to make out the testa in the grain of Maize; we shall learn later (p. 172) that this grain is really a fruit, containing a single seed.

CHAPTER IV

THE GERMINATION OF SEEDS

Most plants produce an enormous number of seeds, which sooner or later fall to the ground. Here they may sprout or germinate, and grow up into mature plants. Soil often contains a great many seeds of various kinds. If, for example, we turn over the soil of a garden in spring or summer, a new crop of weeds soon appears, most of which have sprung from seeds lying in the soil. On one occasion Charles Darwin, the great naturalist, took three table spoonfuls of mud from the edge of a pond, and kept it moist for six months. As seedlings appeared he pulled them up, and found that during the six months no fewer than 537 plants grew from seeds buried in the mud.

Resting period. Most seeds will not germinate immediately they are ripe, for they need a period of rest. As a rule seeds rest during the winter and germinate in the following spring, but many kinds are capable of resting in a dormant condition for a long time, in some cases for 10 or 20 years, or even longer. The popular belief that wheat grains have germinated after resting for thousands of years in Egyptian mummy cases must, however, be regarded as a myth.

At last, when the period of rest is over, the young plant within the seed awakens, as it were, from its long sleep. This re-awakening, during which the embryo once more becomes an active, growing plant, is what we call the germination of the seed.

We must find out, first, what happens when seeds germinate, and secondly, the conditions under which they germinate.

Germination of seed of Broad Bean. Soak some Broad Bean seeds in water for twenty-four hours, and then plant them in damp bog moss or sawdust, in a wooden box. The seeds should be planted in a number of different positions, an inch or two below the surface. Examine some of the seeds at intervals, and find out what is going on.

Soon the testa splits near the micropyle, and the radicle grows out through the opening. Carefully examine a seed at this stage and notice that the splitting has taken place along the edges of the root-pocket, a loose triangular flap being formed, which is

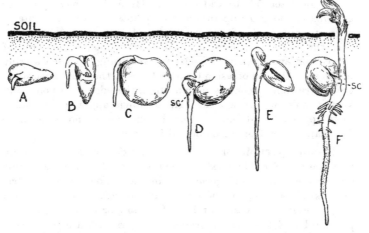

Fig. 21. GERMINATION OF SEED OF BROAD BEAN. In whatever position the seed is sown, the radicle always grows downwards. *sc*, stalk of cotyledon. (× 1/2.)

pushed aside by the growing radicle. As soon as the young root comes out of the seed, it grows straight downwards. It is remarkable that the root always behaves in this way, no matter in what positions the seeds have been planted (Fig. 21). Sometimes the radicle has to bend right round in order to grow downwards (Fig. 21 B).

The testa splits still more, and the stalks of the cotyledons (Figs. 21, 24), which at first are very short, now lengthen, and push the plumule backwards out of the seed (Fig. 21 D). As soon as the plumule is outside the seed it grows upwards, in the opposite direction to that taken by the radicle. The **epicotyl** (Gk. *epi*, upon), which is the internode between the cotyledons and the first leaf above them, is bent into a loop, so that as the young shoot grows, the delicate tip of the plumule is drawn backwards through the soil without being injured (Fig. 21 E). When the plumule reaches the surface of the soil, however, the loop straightens out, and soon the tip of the plumule becomes green. In the meantime the main root, into which the radicle develops, has grown longer, and lateral roots have begun to appear. The two cotyledons remain underground in the seed, gradually withering as the seedling grows.

Germination of Sycamore and Kidney Bean seeds.
Sycamore seeds usually germinate on the surface of the ground,

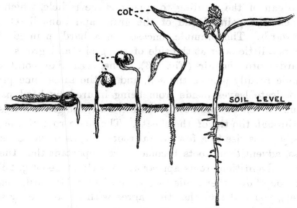

Fig. 22. GERMINATION OF SYCAMORE SEEDS. *t*, testa; *cot*, cotyledon.
(March, × 1/2.)

and in early spring-time you can find great numbers of half-decayed fruits, containing germinating seeds. The fruit splits open and the radicle grows out, easily pushing its way through the thin testa. As soon as it is outside the fruit, the radicle turns downwards and bores its way into the soil (Fig. 22). The **hypocotyl** (the part of the stem below (Gk. *hypo*, under) the cotyledons, between them and the root proper) elongates and pulls the seed out of the fruit, carrying it up into the air. Soon the testa is thrown off and the cotyledons unfold, forming the first green leaves of the seedling plant. Gradually the plumule grows and new foliage leaves develop, the main root in the meantime growing deeper into the soil, and giving rise to lateral roots (Fig. 22).

In the French or Kidney Bean also, the cotyledons come above the surface of the ground and become green. They are dragged backwards out of the soil at the end of a loop, in very much the same way as is the plumule of the Broad Bean.

Germination of Maize. As in the other cases studied, the first organ of the seedling to appear is the radicle, which comes out near the pointed end of the grain, and immediately grows downwards. The plumule, encased in a hard, pointed sheath, appears a little later at the side of the grain, and grows straight upwards into the air (Fig. 23). The spear-like point of the plumule readily pierces the soil, and at the same time prevents the delicate leaves inside from being injured by hard particles of soil. When the plumule reaches the air, the green leaves come out through the tip of the sheath. The main root of the Maize usually gives rise to a few lateral roots but, as in the Cock's-foot Grass, adventitious roots become more important than the main root. Adventitious roots appear at an early stage (Fig. 23).

If you allow other seeds to germinate (e.g. Mustard, Pea, Sunflower), you will find that they agree with those already studied in the following respects :

1. The root always appears first, coming out through the testa at or near the micropyle. It then grows downwards, burying itself more and more deeply in the soil.

2. The young shoot or plumule appears later, and grows upwards towards the light and air.

All germinating seeds behave in this way, no matter in what positions they may have been planted.

The plumule may escape from the seed in several different ways : in the Broad Bean it is pushed out by the growth of the stalks of the cotyledons ; in the Kidney Bean it is pulled out by

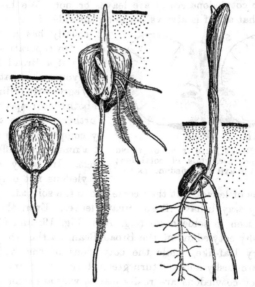

Fig. 23. GERMINATION OF MAIZE. (The two earlier stages ×1; older stage on right × 2/3.)

the growth of the hypocotyl, while in the Maize it frees itself by its own growth.

In most seedlings the cotyledons come above the soil and turn green ; this is known as the **epigeal** (Gk. *epi*, upon, and *ge*, the earth) type of germination ; it occurs in Sycamore, Kidney Bean, Mustard etc. In some seedlings, however, the cotyledons remain underground and so show **hypogeal** (Gk. *hypo*, under) germination, e.g. Broad Bean, Pea, Maize and Oak.

Perhaps you have been wondering why the cotyledons of the embryo are called seed leaves, though their shape is always different from that of the foliage leaves, and in some cases they are not at all like leaves to look at. Now it is very important to cultivate the habit of using words correctly, so we will try to find out whether cotyledons really are leaves or not. We have already learnt that a leaf is always borne on the side of a stem, and that

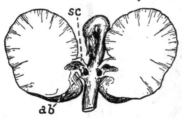

it usually has a bud in its axil. On separating the cotyledons of a Broad Bean seedling, you will see at once that they agree with ordinary leaves in both these respects (Fig. 24). Further evidence is afforded by comparing the cotyledons of a number of different seedlings. In many plants the cotyledons not only have the

Fig. 24. BROAD BEAN, PART OF YOUNG SEEDLING. *sc*, stalk of cotyledon; *ab*, bud in axil of cotyledon. (× 1.)

positions of leaves, but they come above the soil, turn green, and in every way behave like ordinary leaves. Even the veins are clearly seen in some cases (e.g. Ash, Fig. 19, and Castor Oil). The fleshy cotyledons of the Broad Bean, on the other hand, are not very leaf-like, while the cotyledons of the Kidney Bean, which are fleshy and yet turn green, form an intermediate type.

If then cotyledons are really leaves, why are some of them so unlike leaves to look at? Is it because their functions are different from those of ordinary leaves? We shall find out later what the duties or functions of green leaves are, but in the meantime we may learn something of the functions of cotyledons.

We found that Bean cotyledons contain starch, and that they shrivel during germination. Maize endosperm also contains starch, and it too disappears during germination. This suggests that in both cases the starchy food may be used during the growth of the embryo. That this is actually the case is proved by the following experiment.

EXPERIMENT 1. Take ten Broad Bean seeds and allow them to germinate. Then scrape away enough soil to expose five of the seeds, and cut

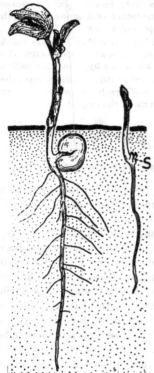

off their cotyledons with a sharp knife, taking great care not to injure the plumules or radicles. Replace the soil and watch the result. The seedlings with cotyledons will grow much faster than those without (Fig. 25). We may therefore conclude that the chief function of fleshy cotyledons, such as those of the Bean, is to store food for the use of the growing seedling during germination.

Try a similar experiment with Maize, in this case removing the endosperm, not the cotyledon. Again the seedlings will grow very slowly, for they have been deprived of the food stored in the endosperm. Under ordinary conditions the embryo absorbs this food through the cotyledon.

In endospermic seeds the chief function of cotyledons is to absorb food (e.g. Maize, Ash), and in non-endospermic seeds to store food (e.g. Bean and Pea).

Fig. 25. BROAD BEAN SEEDLINGS TEN DAYS AFTER REMOVAL OF COTYLEDONS OF PLANT ON RIGHT; AT FIRST BOTH WERE THE SAME SIZE. *s*, stalks of cotyledons. EXP. 1. (× 1/4.)

HINTS FOR PRACTICAL WORK.

1. Take some garden soil and pick out any weeds or pieces of living plants in it. Put it in a pot and water occasionally. Count the seedlings which come up, and see how many kinds there are. Keep a diary with dates when new seedlings appear.

An even better plan is to mark out with pegs a square yard in a garden. Dig up any weeds there may be, and then leave the ground quite untouched for a year or more—the longer the better. Keep a record of the plants which appear, and of what happens to them.

2. Sow several kinds of seeds, such as Mustard, Sunflower, Pea, Marrow, Nasturtium, Wallflower, Onion, Barley. Find out whether germination is hypogeal or epigeal; which organ appears first; how the plumule escapes from the seed, and how it is protected from injury as it grows through the soil; in which directions the radicles and plumules grow.

3. In spring or summer collect different kinds of wild seedlings and make drawings to show the differences between cotyledons and foliage leaves. How many of these seedlings show epigeal and how many hypogeal germination?

4. Cut off the plumules of two or three Broad Bean seedlings when they have come above the soil. Examine the seedlings from time to time, and find out what happens to the buds in the axils of the cotyledons.

CHAPTER V

THE CONDITIONS NECESSARY FOR GERMINATION

We have now found out what happens when seeds germinate; can we find an answer to our second question—under what conditions does germination take place?

We know that seeds will not germinate so long as they are kept dry, but if planted in the moist soil of a garden, they soon begin to sprout. This suggests that water may be necessary. Again, when warm weather comes in spring-time, great numbers of seedling weeds spring up in our gardens, though few appeared during the cold days of winter. Perhaps warmth too has something to do with germination. Seeds usually germinate on or in the soil, so possibly soil is necessary. Finally plants are generally surrounded by air, and at least during the daytime are exposed to light, so it may be worth while to enquire whether air and light help in any way.

We can only decide which of these—water, warmth, soil, air and light—are really necessary for germination, by experiment.

EXP. 2. (a) *Setting up the experiment* (see Fig. 26). Take a number of seeds (Barley grains do nicely); keep a few of them dry, and soak the rest in water for 24 hours. Now take five clean glass jars and put some cotton-wool, or several pieces of clean blotting-paper, at the bottom of each. Place the dry seeds on the cotton-wool in the first jar and divide the soaked seeds equally amongst the others. Boil some water, allow it to cool, and then pour it into jar no. 2, till the seeds are covered to a good depth. As soon as most of the bubbles have risen to the surface, slowly pour a little oil on it. Now pour ordinary water into the other three jars (nos. 3 to 5), till the cotton-wool is thoroughly soaked, but do not cover the seeds with water.

It is obvious that the seeds in jar no. 1 will get plenty of air but no water. Those in jar 2 will have plenty of water, but little or no air, because boiling the water has driven off any dissolved air, and the oil prevents more from entering. The seeds in jars 3, 4 and 5 will get both air and water.

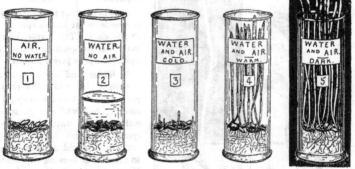

Fig. 26. CONDITIONS NECESSARY FOR GERMINATION OF SEEDS.   EXP. 2.

(b) *Looking after the experiment.* Put jars 1, 2 and 4 in a warm, well-lighted room. Jar 3 must be kept in a cold place, e.g. out of doors (of course sheltered from rain) in the winter, or, in warmer weather, in a tin containing pieces of ice. The ice must be renewed as it melts; it will last longer if you cover the outside of the tin with flannel or felt—can you explain why this is so? Put jar 5 into a dark cupboard, so that no light can reach it.

Jars 1 and 2 will need no further attention, but the other three should be looked at every day, and a little water added when necessary, for the seeds must never be allowed to get dry. If the mouths of the jars are partly covered with glass or cardboard the seeds will not dry up so easily, and dust will be kept out.

(c) *Results of the experiment.* The seeds in jars 1 and 2 do not germinate, neither do those in jar 3, if they have been kept at a very low temperature—if a little warmer, they germinate slowly. In jars 4 and 5 the seedlings have grown much faster than in the cold jar. Those in the light are green and healthy looking, those from the dark cupboard of a sickly yellow colour, though they are even taller than the green ones (Fig. 26).

(d) *What the results of the experiment teach us.* The object of our experiment was to find out whether water, warmth, soil, air and light (or any of them) are necessary for the germination of seeds. Let us see what conclusions we can draw from the results:

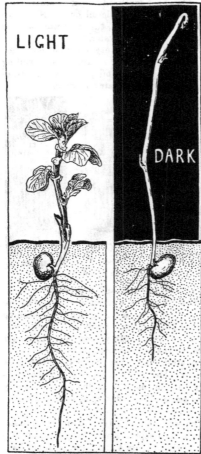

LIGHT

DARK

Fig. 27. EFFECT OF LIGHT AND DARKNESS ON THE GROWTH OF BROAD BEAN SEEDLINGS. EXP. 3. (× 1/5.)

1. Evidently *soil is not necessary*, for the seeds in jars 4 and 5 germinated without it.

2. *Light is not necessary*, for the seeds in jar 5 have sprouted in the dark. The leaves, however, become green only in the light (compare jars 4 and 5).

3. *Water and air are both necessary*, for when the seeds get both (jars 4 and 5) they germinate, but will not do so if either air or water is absent (jars 1 and 2).

N.B. It would not be correct to argue that because dry seeds (jar 1) do not germinate, therefore water is necessary. In order to be able to say definitely that water is necessary, we must prove two things, first that seeds will not germinate without water (jar 1), and secondly that they will germinate with it (jars 4 and 5).

4. *Warmth is necessary*, for if jar 3 is kept at a very low temperature, the seeds will not even begin to germinate, while those at a higher temperature (jar 4) do so.

EXP. 3. Etiolation. To learn the effects of light and darkness respectively on the growth of Broad Bean seedlings, take two flower pots filled with garden soil, and plant four soaked Beans in each. Expose one pot to strong light and keep the other in a dark cupboard; water the plants when necessary. Compare the two sets of seedlings when they have been growing for some time (Fig. 27). The differences between them are even more striking than in the case of the Barley seedlings (Exp. 2). The shoots of the Bean seedlings grown in light are sturdy and green, and have well developed leaves, while those grown in the dark are tall and weak, with long colourless internodes and tiny yellow leaves. The plants grown in the dark are said to be **etiolated** (Fr. *étiolé*, drawn out). If you bring the etiolated seedlings into the light, they will become green, and continue to live, but if still kept in the dark, they will get weaker and weaker and finally die. Evidently then, although seeds do not require light for germination, light is required for the healthy growth of older plants.

Familiar examples of etiolated plants are, grass when covered for some time by a large stone, also "earthed-up" celery and "forced" rhubarb. In the last two cases it is the leaf-stalks which become elongated; in the Broad Bean and many other plants it is the stem, while in the Barley the etiolated leaf-blades are long and narrow (Fig. 26). In the dark then, and even in shady places, plants grow longer and taller than they do in strong light. This property is often useful to plants growing in nature. For instance, seedlings covered by dead leaves or by a thick growth of other plants, may reach the light by means of this rapid growth in length. We shall see later that leaves can only perform their proper functions in light, and it is only after reaching the light that they develop fully.

Respiration of germinating seeds. Why do seeds need air in order to germinate? To answer this question we must first find out whether the air is altered in any way by germinating seeds.

EXP. 4. Take four clean glass jars, with well-fitting corks or stoppers. Put wet cotton-wool or blotting-paper at the bottom of each of the jars, and half fill two of them (jars 1 and 2) with germinating peas or barley. Now cork all four jars very tightly, if necessary using vaseline or plasticine to make them perfectly air-tight, and keep them in a warm place. After two or three days test the air in the jars in the following way. Uncork jar no. 1 very gently and put the lighted end of a taper into it; the taper goes out.

Pour a quantity of clear lime-water into jar no. 2, replace the cork and shake the jar well. The lime-water becomes white and milky, showing that carbonic acid gas (carbon dioxide) is present, for this gas always turns lime-water milky. Test the air in jars 3 and 4 (i.e. those without seeds) in the same way; in these the taper continues to burn and the lime-water does not become milky. Evidently some change has taken place in the air which has been in contact with the germinating seeds. Our lessons in chemistry have taught us that only air which contains oxygen will support combustion. It is clear therefore, that as the taper will not burn in the jars containing the seeds, but will burn in the others, oxygen has been removed from the air in the seed jars. The oxygen could not have escaped from the air-tight jars, neither could it have been removed by the cotton-wool or water, for all the jars contained these; it must therefore have been absorbed by the seeds themselves. For similar reasons the carbon dioxide in jar 2 can only have come from the seeds. We may say then, that germinating seeds take in oxygen and give off carbon dioxide. Now this also happens when we our-selves breathe or respire—we take in air, absorb oxygen from it, and give out carbon dioxide. You can easily prove that carbon dioxide is given off during breathing, by blowing down a glass tube into a beaker of lime-water.

We have now found out that germinating seeds respire like ourselves. Later on we shall study respiration more fully, and shall then learn that the reason why seeds cannot germinate with-out air, is because they need the oxygen in the air for respiration.

When sowing seeds in your garden, you should try to place them under the best conditions for germination, by applying the knowledge gained from your experiments. For instance :

1. Most seeds should be sown in spring; in winter there is too little warmth for them to germinate.

2. The depths at which seeds should be sown depends partly on the amount of stored food, and therefore on the size of the seed. For instance, Broad Beans may be sown 3 inches below the sur-face, Parsnips 1 inch, and very small seeds like Poppies or Onions, $\frac{1}{4}$ inch. If sown too deeply the seedlings may use up all their re-serve food before reaching the surface.

3. Another reason for not sowing too deeply, especially in wet, heavy soil, is that the seeds may not get enough oxygen for respira-tion, for the deeper layers of soil often contain little air and much water (cf. Exp. 2, jar 2)

4. Although seedlings have a remarkable power of forcing their way through the soil, it is well to assist them by sowing in fine, well-worked soil. This not only allows the shoot and root to penetrate the soil easily, but also provides them with plenty of air.

HINTS FOR PRACTICAL WORK.

1. This book is intended to help you to find out things for yourself, not merely to tell you what other people have found out. You should therefore actually perform the experiments described in this and other chapters, always remembering, when you do so, that to make an experiment is really to ask nature a question. If it is worth while to ask the question at all, it is worth while to take the trouble to ask it properly, i.e. in such a way that you can understand the answer which nature gives. Knowledge so gained is really worth having. When trying experiments, always observe the following rules:

(*a*) Be quite clear in your own mind what it is you want to find out.

(*b*) Set up the apparatus carefully, otherwise you cannot trust your results. If your experiment does not succeed, try to find out why it went wrong, and then do it again with greater care.

(*c*) Look after the experiment properly—watering the plants when necessary, and so on.

(*d*) Write an account of the whole experiment, illustrated by sketches of the plants or the apparatus used. Notice exactly what happens, and make full notes of the results.

(*e*) Use your common sense in drawing conclusions from the results.

(*f*) Where possible set up a "control experiment," that is an experiment in which the same apparatus is used as in the main experiment, but some condition or other is different. A control experiment will often help you to check the results of the main experiment, and to draw more correct conclusions. E.g. the two jars without seeds in Exp. 4 form a control experiment; we saw that they helped us to conclude that the seeds themselves (not the cotton-wool or water) absorbed oxygen and gave out carbon dioxide.

2. In Exp. 2 the water for jar no. 2 should be boiled for an hour in a narrow-necked flask, and used as soon as it is cool. A minute quantity of air may get in while setting up the jar, but not enough to interfere with the experiment. Even if a few seeds begin to germinate growth ceases almost immediately.

3. In performing the experiment on etiolation (Exp. 3) examine and draw the specimens carefully; make a full list of all the differences you can find between the specimens grown in light and darkness respectively. Both shoots and roots should be examined.

CHAPTER VI

THE GROWTH OF PLANTS

The embryo of a seed grows into a seedling and the seedling into an adult plant. Annuals like the Groundsel often attain their full size in a few months, while perennials keep on growing for years.

How to measure growth. Does the whole of a plant grow at the same time, or only certain parts of it? This question too may be answered by a simple experiment.

EXP. 5. Take a wide-mouthed, corked bottle and line it with blotting-paper. Pour a quantity of water into the bottom of the bottle, so that the paper may be kept constantly wet. Now choose several Bean seedlings with clean, straight roots about an inch long, and mark the roots in the following way. Bend a piece of copper wire into a loop, and connect the two ends with a piece of thread. Dip the thread into *water-proof* Indian ink, and gently lay it across the root about 1/16 inch behind the tip, so as to leave a fine ink line. Mark the rest of the root with similar lines at equal distances apart (Fig. 28 A). Measure and note the length of each root from its tip to the ink line nearest the seed, and then pin the Beans to the cork of the bottle, so that the roots hang straight downwards. Replace the cork in the bottle and allow the Beans to grow in the dark for a day or two. Then take them out and (1) measure the roots and see how much they have grown; (2) examine the new positions of the ink lines. Near the seed and

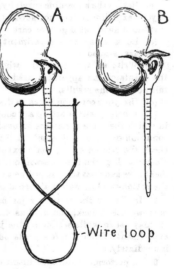

Fig. 28. How to measure growth of Bean roots. Exp. 5.

at the extreme tip of each root, the

lines are in the same positions as before, but those just behind the tip have become widely separated (Fig. 28 B). This proves that *the actively growing region of the root is the part just behind the root-tip.*

Carry out a similar experiment with stems, marking the plumule of the Bean instead of the radicle. In this case the seedlings may be growing in a pot, so the damp bottle is unnecessary. Is there a special region of growth in the stem as well as in the root?

What plants are made of. The food stored in the seed is used up long before the plant is full grown; yet the plant keeps on growing until it is many times larger than it was in the seedling stage. Where has the extra material necessary for all this growth come from? We shall be able to answer this question more easily if we know something about the chemical composition of a plant. You can analyse a plant, very roughly, in the following way.

EXP. 6. Take some small pieces of a plant—any part will do—and heat them over a Bunsen burner, in a thoroughly dry test-tube. Drops of clear liquid soon appear on the cool upper part of the tube; on testing this liquid it is found to be water. In the first place then, the plant contains water. It has been found that in some cases more than nine-tenths of the whole plant consists of water. A given weight of fresh cabbage, for example, contains a little more water and a little less solid matter, than the same weight of milk.

If you now heat the plant more strongly, it becomes black and charred. The black, charred remains of the plant are practically charcoal, an impure form of the chemical element carbon. Finally, if the charcoal is heated still further in a porcelain crucible, it slowly burns away till at last nothing is left but a small quantity of ash.

Some of the materials, then, into which a plant can be split by heat are *water*, *carbon* and *ash*. In order that a plant may grow, or increase in size, it must obtain materials for growth from outside, and these must provide water, carbon, and the substances which form the ash of the plant. We must find out how the plant obtains each of these.

HINTS FOR PRACTICAL WORK.

1. You will find the following useful when performing Exp. 5: Get everything ready before digging up the Bean from the damp sawdust or

moss. Before marking the root, gently remove any moisture from its surface
with torn pieces of blotting-paper. Then mark the root quickly (N.B. mark
as instructed—do not use a pen or ordinary black or red ink!) and put the
Bean into the damp bottle at once. Try to find out why these precautions
are necessary, and remember, in carrying out any experiment, that very
often it is carefulness over the little things which makes all the difference
between success and failure.

2. The damp chamber described on p. 88 (Fig. 57) may be used in Exp. 5
instead of a corked bottle.

3. Young flower stalks of Dandelion are useful for investigating the growth
in length of stems.

CHAPTER VII

ROOTS AND THEIR WORK

How a Plant gets Water and Nourishment from the Soil

The roots of a plant are hidden away underground, in close
contact with the particles of soil; the shoot, on the other hand, is
surrounded on all sides
by air. It seems likely,
therefore, that the food
materials necessary for
growth come either from
the soil or from the air.
We will first try to find
out if the roots obtain
anything from the soil.

If you neglect to water
a pot plant, its leaves
droop and show signs of
withering. If now water
is poured on the soil,
without wetting the
shoot, it soaks into the
soil and the plant re
covers. Apparently water has been absorbed by the roots.

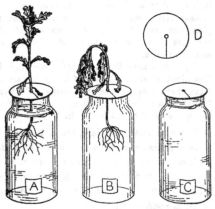

Fig. 29. Absorption of water by roots.
Exp. 7.

EXP. 7. The following is a more exact experiment of a similar kind. Take a clean wide-mouthed bottle and nearly fill it with water. Cut a piece of cardboard a little larger than the mouth of the bottle, make a hole in the centre, and slit the cardboard from the edge to the centre (Fig. 29 D). Dig up a Groundsel plant, shake the roots free from soil, and slide the stem between the edges of the slit in the cardboard, till it occupies the hole in the centre. Arrange the plant as in Fig. 29 A, with its roots in the water. Lastly, pour some olive oil on the surface of the water, and mark the level of the water by gumming a slip of paper on the outside of the glass. Now set up two similar bottles as controls. These should be fitted exactly like the first one (A) except that in one of them (B) the water and oil are left out, so that the roots are in air, the third bottle (C) containing water and oil, but no plant.

Examine the experiment every day and notice what happens. The plant without water (B) soon withers, but the one with water (A) remains fresh, and the level of the water sinks slowly day by day. Evidently water is being absorbed by the roots, for the water level in (C) remains the same, the oil preventing evaporation from the surface in both cases. We may conclude from this experiment that an important function of roots is to absorb water from the soil.

Parts of a root. Now we must study the roots themselves a little more closely. Allow some Barley grains or Cress seeds to germinate on damp blotting-paper in-side a biscuit tin. Examine some of the young roots (Fig. 30) with a lens and notice (*a*) the tip, covered by a minute cap, the **root-cap.** The edges of the root-cap generally fit so tightly round the root, that it is difficult to see, unless, as is sometimes the case, the cap is slightly different in colour from the rest of the root. Hyacinth roots growing in water in a glass often show root-caps very distinctly. The duty of the root-cap is to protect the

Fig. 30. ADVENTITIOUS ROOTS OF BARLEY. *c*, root-cap; *gr*, growing region; *h*, root-hairs. (× 3.)

delicate growing tip of the root from injury as it pushes its way amongst the hard particles of soil. (*b*) The bare **growing region** just behind the root-cap (see Exp. 5, p. 36), and (*c*) still nearer the seed, a large number of extremely delicate white hairs, covering the surface of the root. These **root-hairs**, as they are called, are never found on the growing region itself, but always behind it. If they did occur on the growing part, they would be broken off as the root forced its way through the soil. Almost all roots, even the smallest branches, show these three parts— *root-cap, growing region,* and the *region of root-hairs* (Figs. 23, 30)—though in some cases the hairs are so small as to be scarcely visible. As the root gets older and longer, the root-hairs die away, only to be replaced by new crops nearer the tip, but always just behind the growing region. Lateral branch roots may now appear on the older parts of the main root (Fig. 21 F).

How roots absorb water. In many roots the older parts become hard and corky, the younger parts near the tip, and especially the root-hair region, being the only parts through which water can be absorbed. The root-hairs are indeed of great importance for the well-being of the plant, for they are the chief water-absorbing parts of the root. The exact process by which root-hairs take in water from the soil is rather complicated. and need not be fully described in this book. Briefly it is as follows : Each root-hair is a minute hollow tube, containing a liquid called the sap, which has the power of attracting water from the soil. There are no visible openings anywhere in the tube, so that the only way water can pass in is through the walls of the tube. This is not a difficult matter, for the walls are porous, being composed of a substance which allows water to soak through in somewhat the same way as soft paper or thin, unglazed earthenware would do. Large quantities of water are taken in by the roots in this way, and quickly travel to other parts of the plant.

How roots obtain mineral nourishment. If a Groundsel plant is kept with its roots in pure (distilled) water, it may live for a considerable time, but will probably not grow very much. If, however, a little soil is mixed with the water, the plant will grow and flourish. Evidently then the plant gets something besides water from the soil. If we put some sugar or salt into water, it is dissolved by the water and disappears. The liquid is now said to be a solution of sugar or salt, as the case may be. If the solution is put into a clean watch glass, and allowed to dry up, or evaporate, the sugar or salt remains behind, and can be recovered in a solid form.

EXP. 8. Add some soil to pure water, shake well and allow to stand for some time; then filter the liquid and evaporate it to dryness. A little solid substance is left behind. The water in the soil which is absorbed by the roots of the plants is really not pure water, but a very weak solution of various mineral substances, or salts as chemists call them. These salts are the substances from which the "ash" of the plant is derived (see Exp. 6).

We have now learnt that:

1. Water and mineral salts are necessary for the well-being of plants.

2. These substances are obtained from the soil by the roots of the plant.

3. The root-hairs are the chief absorbing parts of the root.

The question of how the plant obtains its carbon must be left till a later chapter.

Two kinds of root-system. In many plants the radicle of the embryo becomes the **main** or **tap-root** of the adult plant. Lateral branches arise from the tap-root, and these in turn may give off still smaller rootlets, the main root, however, remaining larger and more important than any of the branches. Such a system is called a **tap-root-system,** e.g. Broad Bean (Fig. 27) or Parsnip. In other plants the radicle does not develop into a main root, the root-system being made up of fibre-like **adventitious** roots (p. 11), arising from the base of the stem. This is called a **fibrous root-system** (Fig. 9). The Groundsel sometimes forms a more or less intermediate type; Fig. 1 shows a small tap-root, with rather large lateral branches, and adventitious roots arising from the base of the stem, amongst the lowest leaves.

Lateral roots, whether they arise from another root, or adventitiously from a stem, are always formed inside the parent organ, and push their way to the outside as they grow. If you look at the outside of a Broad Bean root, you can see little slits where the branch roots have come out (Fig. 31).

Fig. 31. LATERAL ROOTS OF BROAD BEAN ARISING INSIDE TAP-ROOT. A, part of tap-root seen from outside. B, tap-root cut vertically through the middle. (× 3.)

Arrangement of the roots in the soil. Roots have a general tendency to grow downwards (p. 26), and this is why they are as a rule underground organs. But all roots do not grow in the same direction. In a tap-root-system the main root grows vertically downwards, like the radicle from which it was developed. The lateral roots, on the other hand, spread out in various directions round the main root, and grow obliquely downwards (Fig. 27). The smallest rootlets grow in all sorts of directions. In a fibrous root-system too, the adventitious roots tend to spread themselves out, and grow obliquely downwards. You will see at once that this spreading out of roots in the soil has two advantages. (1) It enables the root-system to tap the supplies of water in the soil very effectively. (2) It fixes or anchors the plant firmly to the soil, so that it is not easily uprooted by wind or other agencies.

We are now in a position to distinguish clearly between what a root is, and what a root does.

I. *What a root is.*

(*a*) The main root is continuous with the main stem, but lateral roots arise as little growths inside a larger root or, in the case of adventitious roots, inside the stem.

(*b*) The tip of a root is covered by a root-cap.

(*c*) A root does not bear leaves or flowers, and is therefore not divided into nodes and internodes.

II. *What a root does.*

A root has two main functions :

(*a*) To absorb water and mineral nourishment from the soil, in most cases the root-hairs being the chief absorbing parts.

(*b*) To anchor the plant firmly to the soil.

In some cases a root may perform other duties, e.g. storing food (p. 115), climbing (p. 77).

HINTS FOR PRACTICAL WORK.

1. In Exp. 7, enough oil should be used completely to cover the surface of the water. Pour it on to the water very slowly and carefully *after the plant is in position.*

2. Germinate some Barley grains or Cress seeds for yourself. Draw one of the seedlings, labelling the different parts of the root.

3. Dig up a few common plants, e.g. Dandelion, Creeping Buttercup, Chervil, Grass, noticing as you do so, how the roots are arranged in the soil. What kinds of root-system do you find in these plants? Draw the root-system in each case.

4. Dig up a large Groundsel plant without injuring the roots. Wash in water, and then measure the total length of the root-system. This may be done by cutting off all the roots at the base, and arranging them in little piles according to length. Measure the length of one root from each pile, and multiply this by the number of roots in the pile. Finally add all the results together, this will give you the approximate length of the root-system. You may be surprised at its length as compared with the height of the plant.

CHAPTER VIII

THE SOIL

What soil is made of. We shall understand the work of roots better if we know something about the soil in which they grow. To find out what soil is made of, we can experiment with ordinary garden soil in the following ways:

EXP. 9. (a) **Solid matter.** Mix some soil into a paste with water, taking care to break up all the lumps. Put the paste into a bottle, nearly fill the bottle with water and cork it up. Shake thoroughly for some time, then put the bottle where it will not be disturbed, and watch what happens. As the soil settles, you will see (a lens is useful) that the particles become arranged in layers according to size, the heavier ones (small stones or sand) at the bottom, and the finer, more muddy particles at the top. The water may perhaps remain cloudy for days or even weeks showing that some of the finest particles of all are still suspended in it.

Some solid particles, however, have risen to the surface instead of sinking to the bottom. These are usually black in colour, and some you may recognize as decayed fragments of plants. Similar black, decayed matter is found in most soils, and is called **humus.** Humus, like any other organic matter, can be burnt, and if you heat soil very strongly (cf. Exp. 6), the humus burns away, leaving only the mineral particles, which will not burn.

(b) **Water.** Weigh about 100 grams of fresh soil, spread it out on paper in the lid of a biscuit tin and leave it to dry for a few days. To complete the drying process the lid may be put into an oven for an hour or two. Now weigh the soil again being careful not to leave any behind on the paper. The soil will feel dry and will have lost weight, the loss being due to the evaporation of water from the soil.

(c) **Air.** If you plunge a flower pot full of soil into water, bubbles rise from the surface. Evidently as the water soaks into the soil it drives air out. Soil therefore contains air.

We have now found out that soil is made up of (a) solid matter, consisting partly of mineral particles of various sizes, and partly of decaying organic matter or humus, (b) water and (c) air. As a rule the water in soil forms thin films round the solid particles, the air occupying the spaces between them.

Movements of water in soil. These movements may be investigated by means of the following experiments :

EXP. 10. Fit a perforated disc of tin into a glass funnel. Put three or four table spoonfuls of well dried soil on the tin, and pour two table spoonfuls of water on the soil. Measure what runs through; it will always be less than the amount used. Water then tends to drain away downwards, but some is left behind in the soil.

EXP. 11. Tie some muslin over one end of a wide glass tube (a tall lamp chimney will do), and fill the tube with dried soil, packing the particles tightly together. Sow some cress seeds at the top, and stand the muslin-covered bottom of the tube in water. The water gradually rises in the soil, always moving from wetter to drier parts, till at last it reaches the seeds, which begin to germinate.

Water therefore can move upwards in the soil as well as downwards, and from a wet to a dry part. This circulation of water in the soil is of great importance to plants, for it helps to keep up a constant supply to the roots. After rain, the water supply in the soil is replenished, the surplus draining away downwards. During drought, the upward movement brings some of the water from deeper levels once more within reach of the roots.

How soil has been made. The exposed surfaces of rocks are continually being worn away and broken into fragments by the action of wind, rain, frost, snow and other agencies. This process, often an extremely slow one, is called the **weathering** of rocks. The broken fragments are again weathered into still smaller pieces till finally they become the mineral particles of which soil is largely composed. Sometimes these particles accumulate where they are formed, gradually covering the solid rock with a layer of soil. At other times they are carried away by running water, wind, or even moving ice, giving rise to soils at a distance. **Alluvial soils** are formed by running water ; they are chiefly found along river valleys. Usually the water has roughly sorted out the particles according to size, for the coarser ones sink to the bottom first, and the finer ones later (cf. Exp. 9 a). Most of our soils, however, are **drift soils,** which were carried by moving ice, during what is known as the Great Ice Age.

Kinds of soil. Different kinds of soil may be formed either from different kinds of rock, or by the sorting out according to size of particles derived from the same rocks. **Sandy soils** are composed of loose, coarse particles or grains. They dry quickly after rain, and are light and easy to cultivate. On the other hand, **clayey soils** are made up of very fine particles, through which water does not easily drain away. They are heavy and difficult to cultivate, being hard in dry weather and sticky in wet weather. Most of our best soils are **loams,** which come between sands and clay; they contain a good mixture of coarse and fine particles. If the soil contains a very large amount of lime, it is said to be **chalky.** You should repeat Exps. 9, 10 and 11, using different kinds of soil, and find out for yourself which soils retain most water, and which allow it to move upwards and downwards most freely.

The upper layers of soil (i.e. the soil proper) are nearly always darker in colour than the lower layers, which form the **subsoil.** This is because the upper layers are enriched by humus derived from decaying roots, leaves and other parts of plants. Humus is important because, amongst other things, it helps to keep water in the soil. Some soils contain much more humus than others, for instance, in a wood the surface layers of soil are often very rich in humus ("leaf mould"). Again, the **peat** of our bogs and moorlands contains very little mineral matter, for it is chiefly composed of the dead remains of countless generations of plants[1].

Cultivated soil. Farmers and gardeners plough or dig the soil in order to break it up, improve the drainage, and allow air and warmth to enter. They know that if the soil is too compact or too wet, their crops will suffer. Yet wild plants flourish without the help of man! In nature the work of cultivating the soil is largely performed by burrowing animals, especially earthworms, as Mr Charles Darwin showed many years ago. These industrious

[1] A great deal of useful information about soils is found in Sir E. J. Russell's *Lessons on Soil,* Camb. Univ. Press.

little animals make innumerable tunnels in the soil, often to a
depth of 5 or 6 feet. Worm burrows allow air to enter the soil,
improve the drainage, and form passages along which roots can
easily grow.

HINTS FOR PRACTICAL WORK.

1. Carefully record the results of your experiments with different kinds of
soils, such as sand, clay, loam, chalk, peat. For instance, measure and
record (a) the actual amounts of water which run through the different soils
(Exp. 10), using the same weight of soil and volume of water in each case;
and (b) the heights to which the water rises in the different soils in a given
time (Exp. 11), and so on.

2. Compare the results of your experiments, and write an account of the
properties of the different soils you have examined. In which of them would
you think an ordinary plant ought to flourish best? Give reasons for your
conclusions.

CHAPTER IX

THE TRANSPIRATION OF WATER

The water channels. Where does the water go after it
has been absorbed by the roots? To answer this question several
experiments will be necessary.

EXP. 12. Take a Groundsel or other small plant, cut off the tips of the
larger roots and stand the roots in water coloured with red ink. After a time
the veins of the leaves become tinged with red. If you split open the stem
you will also find streaks of red running up inside it; these streaks are the
veins of the stem. The coloured water has evidently been taken in and
carried or conducted from the roots to the leaves. Only the veins are coloured
red, so we may conclude that they are the channels along which the water
travels.

A good way of examining the veins is to cut off a Plantain or
Primrose leaf close to the stem, putting the cut end into red
ink and water as before. After a few hours snap the leaf across
(towards the lower end) by bending it sharply backwards and
forwards. Then gently separate the two parts of the leaf; the

veins being tough, some of them will be pulled out and appear as red strands. Each vein contains a bundle of extremely fine, delicate tubes, the **water-vessels** or -pipes. They may be compared with the water-pipes of a city, which carry water from the reservoir and distribute it to the houses in every part of the city.

What becomes of the water. If we leave a shallow dish of water on a table, the water gradually evaporates and dries up. Similarly, a plant left without water withers and dries up (Fig. 29). This suggests that plants may give off water by evaporation.

EXP. 13. To find out if this is really so, fit a dry, wide-necked bottle over a young growing plant as in Fig. 32. Make a hole in the cork with a cork borer, a little larger than the stem of the plant, and then cut the cork in two vertically (Fig. 32 A, B). Fit the cork into the bottle so that it clasps the stem, making all joints air-tight with plasticine. Soon the inside of the bottle becomes misty, and before long drops of water appear on the glass. As the bottle is air-tight, this water must have come from the shoot of the plant itself. Further, it has been evaporated, that is, given off as water-vapour, for it only becomes visible when it condenses on the cold glass. Arrange a control experiment without the plant, and see if water appears in the second bottle.

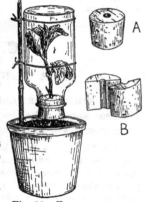

This process, by which water is evaporated from a living plant, is known as **transpiration**.

Fig. 32. Transpiration of
WATER. Exp. 13.

EXP. 14. To find out how much water is transpired, take a plant growing in a small pot. Water the plant and then put the pot into a thin glass jam jar, arranging the top of the pot on a level with the top of the jar. Cut a piece of American cloth in the same way as the cardboard in Exp. 7 (Fig. 29), and tie it tightly over the mouth of the jar, leaving the shoot of the plant outside. Seal any places where water-vapour might escape with plasticine or wax. The roots and soil are now enclosed in a nearly air-tight case, so any water which is lost must be transpired from the shoot. Put the pot on a pair of scales and weigh it carefully. By the next day it will be found to have lost weight, owing to water having been transpired. In this way you can find out how much water is lost by transpiration in a given time. A control might be set up with a glass rod instead of the plant.

Conditions which increase or lessen transpiration. You know that wet clothes hung out in a garden dry quickly on a hot or windy day, but slowly on a cold or still day. This is because the rate at which water evaporates depends on the condition of the air. Experiment 14 may be used to find out the conditions which increase or lessen transpiration. Keep the pot in a draught for two hours, carefully weighing it at the beginning and end of this time. Much more water will have been lost than if the plant had been sheltered from wind under a bell jar. Similarly, more water is transpired in sunshine than in shade, and on a dry day than on a rainy one. *Transpiration* is, indeed, affected by just those conditions which affect the evaporation of water from wet clothes, i.e. it *is increased by heat, wind and dry air, and lessened by cold, and still or damp air.*

How the water escapes. The shoot of the plant is covered with a thin skin, which is nearly waterproof, and yet the water manages to escape. It does so through little pores or openings known as **stomata** (Gk. *stoma*, a mouth). These stomata are very numerous, but so minute that you cannot see them even with a pocket lens. In most cases the stomata are found chiefly on the under side of the leaf.

EXP. 15. The presence of pores in the skin of a leaf may be shown by sealing up the cut end of the petiole of an uninjured Lilac or other leaf, and then plunging it into hot water. The heat causes the air in the leaf to expand, and the surface becomes dotted with numerous air-bubbles, which have been forced out of the stomata.

Why plants transpire. The amount of water transpired is often very great; a large Sunflower plant, for instance, may give off a pint of water in the form of vapour during a single dry day in summer. Indeed, by far the greater part of the water taken in by the roots is transpired by the leaves, only a very small part being kept by the plant. At first sight it may appear unnecessary for the plant to absorb so much water, only to give off most of it again unchanged. But we must remember that the mineral salts which the plant needs are present in the soil-water only in very small quantities. Hence in order to get enough of these salts, the plant must absorb far more water than it actually requires. The plant acts like a kind of filter, keeping the salts, but allowing most of the water to escape.

We have now seen that there are three important processes connected with the water supply of the plant.

1. The taking in of water **(absorption)** by the roots and root-hairs.

2. The carrying of water **(conduction)** from the roots to other parts of the plant by the veins.

3. The giving off of water **(transpiration)** from the leaves through the stomata.

It is of the greatest importance that these three processes or functions should be properly balanced. If, for instance, during a spell of hot, dry weather, the plant were to give off much more water than it absorbed, it would gradually dry up and die (cf. Fig. 29). Sometimes this actually happens, but not very often, for, as we shall see in the next chapter, plants have various ways of economizing water.

HINTS FOR PRACTICAL WORK.

1. Many of the experiments in this book may be done in different ways. For example, a striking form of Exp. 12 is to use a cut shoot of a plant with white flowers (e.g. White Phlox or Narcissus). When the veins of the petals become red, they show up clearly against the white background of the petals.

The reason for cutting off the tips of the roots in Exp. 12 is to allow the red colouring matter to get into the veins.

2. In Exp. 14, tinfoil or other waterproof material may be used instead of American cloth. Again, instead of the flower pot and jam jar, you may use a tobacco tin of the "lever lid" kind, the tin lid being replaced by a split cork prepared as in Exp. 13. Take the plant carefully (with the soil) out of its pot and put it into the tin, so that the stem passes through the hole in the cork. Make the joints air-tight as in Exp. 13.

CHAPTER X

HOW PLANTS ECONOMIZE WATER

You may have noticed that on a hot summer day plants often droop for want of water, but that during the night they recover again. Plants do not transpire much at night, partly because the air is cooler and moister (cf. p. 49), and partly because the stomata close in darkness. The roots, however, absorb water by night as well as by day, so that the plant can usually make up at night-time for water lost during the day. But in some cases this is not sufficient, and we find that plants living in soils which contain little water, or in very hot or windy places, generally have special ways of lessening transpiration and of economizing water.

Small leaved plants. Your experiments on soils (Chap. VIII) will have taught you that clays and loams contain far more water than sandy soils. The difference is so great that many wild plants which flourish on an ordinary soil cannot grow on a dry, sandy one. Many plants found on clay have large leaves, e.g. Oak trees, Brambles, Primroses and Foxgloves. Even the grasses have broader leaves (e.g. Cock's-foot, Figs. 9, 10) than those which grow on dry soils. On the other hand, many of the plants of a sandy heath or other dry soil have comparatively small, narrow leaves, e.g. Pine trees (Fig. 38 B), Heather and Ling (Fig. 33 A), Gorse (Fig. 34). The grasses too, such as Sheep's Fescue, often have narrow leaves (Fig. 33 c). Even when the same kind of plant can grow on both damp and dry soils, the leaves of the plants from dry soil are usually smaller than those from damper soil. A Groundsel plant on a dry gravel path, for example, is much smaller than one from a rich, moist garden soil. Small leaves give off less water than large ones, so plants with small leaves are generally better able to live in dry places than are those

4—2

with large leaves. We must now examine a few plants living in dry places a little more closely.

Plants with rolled leaves. Ling (Fig. 33 A) is found on dry, sandy heaths and on moorlands, where it is exposed to strong, drying winds. Its leaves are very small and crowded together, and on the under side of each is a groove produced by the rolling

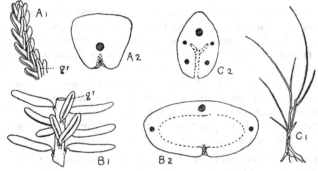

Fig. 33. "Rolled" leaves. A, Ling. B, Crowberry. C, Sheep's Fescue grass. *gr*, groove on under side of leaf. A 2, B 2, C 2 are cross sections of the three leaves; the stomata open along the dotted lines; shaded patches are veins. (A 1, B 1 × 3; C 1 × 3/4; A 2, C 2 × 30; B 2 × 20.)

or curling of the leaf. The stomata, through which water-vapour escapes, are well protected from the drying action of wind, for they are inside the groove, covered up by numerous hairs. Similar "rolled" leaves are found in other moorland or heath plants, e.g. Heath, Crowberry and Sheep's Fescue grass (Fig. 33 B, C). The Marram grass of our sand-dunes has leaves which roll up during drought and unroll when water is plentiful.

Spiny plants. Gorse or Furze is a familiar evergreen plant which often grows on heaths or on windy mountain slopes. Its branches form sharp, pointed spines. The leaves too are narrow and spiny, and there is little surface from which water can be transpired (Fig. 34); this helps to fit the plant for life in dry or windy places. At first sight it is not easy to distinguish the leaves from the smaller stems or branches. The way to do so is to notice carefully their positions on the parent stem, remembering that branches arise from buds in the axils of leaves. Fig. 34 B shows a single long spine, which must be a branch because it is borne in the axil of a leaf. There are several pairs of smaller spines on the long one, each pair consisting of a leaf with a small branch stem (nearer the tip) in its axil.

The lowest leaves of Gorse seedlings are quite unlike those of the older plants, their shape being more like that of Clover, which belongs to the same family (cf. Figs. 35 and 52).

The Hawthorn has two kinds of spines, both of

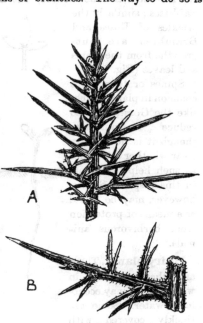

Fig. 34. Twigs of Gorse. (Sept. A × 1½; B × 2¼.)

which are branch stems, for they too arise in the axils of leaves, and, like other branches, may bear leaves and buds. The smaller spines (Fig. 36 A, B) bear minute scale leaves when young. These scales soon fall off, but the following year one or two buds, in the axils of the lowest scales, may develop into dwarf branches. The larger spines (C) bear ordinary foliage leaves, and as a rule most of the axillary buds on the spine develop into dwarf branches. In the Holly (D) and Thistle the foliage leaves themselves are spiny, while in the Barberry the leaves on the longer branches are entirely reduced to spines. The prickles of the Gooseberry (E) are neither stems nor leaves, but merely outgrowths from the leaf-bases; similarly the prickles of Roses and Brambles are out-growths from the stems and leaves (Fig. 51).

Spines or thorns are common in plants which, like the Gorse, need to reduce transpiration, though it is not always clear whether the spines as such help the plants in this respect. Spines, however, may be useful as a means of protection from herbivorous animals.

Hairy plants. There are other ways, too, in which plants may economize water. Some are thickly covered with hairs which, like the grooves of rolled leaves, protect the plant from the drying action of

Fig. 35. GORSE SEEDLINGS. (× 3/4.) A, a lower, B, an upper leaf. (× 1½.) (Sept.)

wind. The leaves of the Mullein are entirely covered with hairs, while those of the Meadow Sweet have hairs only on the under side, where the stomata are.

Plants which store water. Plants which live in rocky places, where there is little soil to hold water, often have thick, fleshy leaves. These plants absorb water during rain and store it

up in their leaves for the dry time which is sure to come. The
Yellow Stone-crop (found on walls and sandy soil), House Leek
(roofs of houses), and Cactus (American deserts) are examples of
water-storing plants. In Cacti the water is stored in fleshy stems,
not in leaves.

Fig. 36. SPINY PLANTS. A, B, C, Hawthorn. D, Holly. E, Gooseberry. (B, May, × 1. The rest, Sept. × 1/2.)

Habitats. Plants are found in many different situations or
habitats. The kind of soil, the amount of available water, and
other conditions vary very much in these habitats, and what suits
one kind of plant may not suit another. Hence we find that each
distinct kind of habitat, e.g. woodland, moorland, sand-dune, sandy
heath, water, has to a large extent different kinds of plants
growing in it. In some habitats there is plenty, in others a
scarcity of water. In the latter, i.e. in hot or windy places, or
on dry soils, where transpiration might be excessive, only those
plants can flourish which are able to economize water. We have
seen that this can be done in different ways, e.g. by (1) producing
very small leaves, (2) protecting the stomata from wind, either by
"rolled leaves" or by a covering of hairs, or (3) the storing of
water in fleshy leaves or stems.

HINTS FOR PRACTICAL WORK.

1. Try to find as many of the plants mentioned in this chapter as you
can; examine them carefully, and make drawings to illustrate the points
referred to.

2. Collect leaves of typical plants from (*a*) a sandy heath, (*b*) a meadow and hedgerow on clay soil, (*c*) a dry moorland, and (*d*) the undergrowth in a damp wood. Trace the outline of one leaf of each kind on squared paper, and find the sizes of the leaves by counting the number of squares in the various tracings. In this way you can calculate the average size of leaf for the different habitats. Is it true that on the whole the leaves from dry habitats are smaller than those from damp ones?

3. Make a list of all the different kinds of hairy and spiny plants you find during your walks. Note whether the hairy leaves have hairs on the upper or on the under side or on both. Try to find out the nature of the spines of the spiny plants. A special note-book might be kept for "field notes" of this kind.

CHAPTER XI

HOW A GREEN LEAF MAKES FOOD FROM THE AIR

The making of starch. We have seen that a plant obtains from the soil the water and mineral salts it needs. But **carbon** is also necessary for growth (p. 37), and we must try to find out the source from which the plant gets carbon. We may first make two or three simple experiments.

EXP. 16. Put two similar potted plants, e.g. Geraniums, into a dark cupboard for two or three days. Then expose one (label it A) to sunlight for four or five hours, keeping the other (B) still in the dark. Take a leaf from each plant, put both leaves into boiling water for a minute or two, and then soak them in methylated spirit. Gradually the leaf becomes bleached and white, for the green colouring matter soaks out into the spirit. So far we can see no difference between the two leaves, but if they are now dipped into a weak solution of iodine, a striking difference is seen. The leaf from the dark cupboard (B) undergoes no change, but that from the light (A) turns a bluish-black colour, showing the presence of starch (see p. 19). Now change the positions of the two plants, putting A into the dark and B into the light. After a few days test another leaf from each plant with iodine. The starch has disappeared from plant A but is present in plant B.

EXP. 17. Next take a leaf from some variegated plant (e.g. variegated Maple, Fuchsia or Geranium) which has been exposed to sunlight for some hours. Make an accurate drawing of the leaf, showing the positions of the green and colourless parts. Then bleach the leaf and apply the iodine test

as in Exp. 16. On comparing your drawing with the leaf itself, you will see
that there is starch in the parts
of the leaf that were formerly
green, but not in the other parts
(Fig. 37).

EXP. 18. Heat some starch in
a test-tube. It becomes black and
charred, showing that it contains
carbon (cf. Exp. 6, p. 37).

We are now in a position
to draw several conclusions
from our experiments: (1) that
leaves can make starch, a food
(p. 19) which contains carbon,
(2) that leaves can only make
starch if supplied with light,

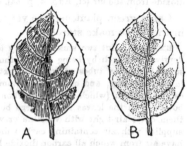

Fig. 37. LEAVES OF VARIEGATED FUCHSIA.
A, fresh leaf, green parts shaded.
B, result of iodine test, dotted parts
contain starch. (× 3/4.) EXP. 17.

and if green colouring matter is present, (3) that the starch
disappears from the leaves in darkness.

The source of carbon. But where has the carbon in the
starch come from? Carbon itself is a solid substance; the greater
part of charcoal, for instance, consists of solid black carbon
(p. 37). Many chemical compounds which contain carbon are
also solids (e.g. starch). Others, however, are liquids (e.g. petrol)
or even gases (e.g. carbon dioxide or carbonic acid gas). Now
it has been found by experiment that if the necessary water and
mineral salts are present, a plant may grow quite well in soil, or
even in water, which contains no carbon compounds at all. Yet
as it grows the amount of carbon in the plant increases. In this
case the only other possible source of carbon is the air. Air is a
mixture of various gases, about one-fifth being oxygen, and nearly
four-fifths nitrogen. It also contains some water-vapour, and a
very small proportion (about 3 parts in 10,000 of air) of carbon
dioxide.

EXP. 19. To prove that air contains carbon dioxide, pour a little clear
lime-water into a clean watch glass, and expose it to the air. Soon the

surface of the liquid turns milky, showing that it has absorbed carbon dioxide from the air (cf. Exp. 4, p. 34).

Can a green plant use the very small amount of carbon dioxide in the air to make starch?

EXP. 20. Cut two small, healthy leaves from a plant which has been deprived of starch by being kept in the dark, and put their stalks into two small bottles or tubes of water. Next take two wide-necked bottles, *the larger the better*, and cover the bottom of one of them with a solution of caustic potash (which absorbs carbon dioxide). Now suspend the small bottles with the leaves in the large bottles, corking the latter and making them quite air-tight with vaseline or wax. The leaf in one bottle will be supplied with air containing carbon dioxide, but the leaf in the other will have air from which all carbon dioxide has been removed by potash. Expose the bottles to sunlight for a few hours, then bleach both leaves and soak in iodine as in Exp. 16. The leaf supplied with carbon dioxide will turn bluish-black, the other being unchanged. This experiment proves that the carbon dioxide of the air can be used by a leaf to make starch. N.B. For this experiment to succeed, it is necessary that the leaves should be very small as compared with the size of the bottles; can you explain why?

How green plants affect the air. One more experiment is needed to find what effect a green plant, when exposed to light, has on the air around it.

EXP. 21. Pour a little water into two wide gas jars, and in one of them (A) place a number of healthy green shoots, with a lighted candle projecting above them. Slip a greased plate over the mouth of the jar, so as to make it quite air-tight. Set up the second jar (B) in exactly the same way, but using dead green shoots (killed by boiling a minute in water) instead of living shoots. When a candle burns, it uses up oxygen and gives off carbon dioxide, and as soon as all the oxygen in the jars has been used, the candles will go out (cf. p. 34). The air in both jars is now very poor in oxygen and rich in carbon dioxide. Expose the jars to bright sunlight for a few hours, then partially uncover jar (A) and introduce a lighted taper. The taper continues to burn, showing that oxygen has been added to the air, and if we test with lime-water, we find that the air now contains little or no carbon dioxide. If the second jar is treated in the same way, the taper goes out and the lime-water rapidly becomes milky.

We see therefore that when exposed to light, living green plants alter the composition of the air around them, taking carbon dioxide from it and adding oxygen to it.

Photosynthesis. The results of the six experiments in this chapter have told us a good deal about the process by which a green plant makes carbon-containing food, without which it can neither live nor grow. For many years botanists have been trying to find out all about this process, and the main facts are now well known, though some things still remain to be discovered. We know, for instance, that carbon dioxide enters the leaf, just as water-vapour leaves it, through the stomata. In the leaf the carbon dioxide is split up; the carbon combines with water to form starch (or in some cases sugar), while the oxygen is set free. Starch and sugar are complex foods which are built up, through the agency of light, from the simple compounds carbon dioxide and water. The process is known as **photosynthesis** (Gk. *phos*, light, *synthesis*, a putting together) or **carbon assimilation.**

Light and chlorophyll. But why are light and the green colouring matter necessary (Exps. 16, 17)? Because the process by which carbon dioxide is split up, and the carbon combined with water, is a chemical one requiring a good deal of **power** or **"energy."** You know from your lessons in chemistry that water is a liquid formed from the two gases hydrogen and oxygen. But we do not get water simply by mixing hydrogen and oxygen, even if we mix them in the right proportions. If however, the mixture is supplied with "energy" in the form of an electric spark, the gases combine instantaneously, and water appears. In the same way, a mixture of carbon dioxide and water does not make starch. But in the green leaf, the power or energy of the sun's rays (i.e. light) causes the two substances to combine and form starch, the superfluous oxygen being set free.

The green colouring matter of plants is called **chlorophyll** (Gk. *chloros*, green, *phyllon*, leaf). Mount a Moss leaf, one of the thin, transparent kind, in a drop of water, and examine it under a microscope. You will see that the chlorophyll is contained in minute, rounded granules. Similar chlorophyll granules occur in vast numbers in all green parts of plants. It has been discovered that the part played by chlorophyll in photosynthesis is to act as a kind of trap for light. It absorbs some of the sun's rays (i.e. light energy), and keeps them in the leaf, where they are used in the chemical process of making starch. Chlorophyll itself is only formed in light (Exps. 2, 3), and when formed, its function is to absorb light.

Perhaps the whole process is most easily understood if we look upon a green leaf as a kind of chemical factory in which food (starch or sugar) is manufactured. In this leaf-factory:

1. The raw materials used are carbon dioxide and water.

2. The chief manufactured product is starch or sugar.

3. The "by-product" formed during the process is oxygen.

4. In an ordinary factory the power or energy necessary to drive the machinery is usually derived from the burning of coal; in the leaf-factory it is supplied by the light of the sun.

We have yet to study some of the ways in which leaves are able to obtain the light they need for photosynthesis; but first we must study the leaves themselves a little more closely.

HINTS FOR PRACTICAL WORK.

1. In performing experiments such as No. 20, you must be very careful to see that the joints are really air-tight. Often the success of the whole experiment depends on this. Use only sound corks; rubber stoppers are the best, but expensive. Perfectly air-tight joints can usually be made between rubber and glass, without the aid of wax. But if ordinary corks are used, any small holes should be filled with vaseline or wax, and a little of the same substance worked in round the joints. Do not use too much vaseline, it only makes things messy. A good wax mixture (which will keep for a long time) is made by melting 35 parts of beeswax with 50 parts of vaseline, and, while the mixture is hot, stirring in 15 parts of powdered resin.

2. When setting up an experiment, always see that your apparatus is perfectly clean.

3. A modification of Exp. 16 is to darken part of a leaf, instead of the whole plant. Cut two circular pieces from the end of a bottle cork, and pin them on either side of a leaf, opposite to each other. Expose the leaf to sunlight and test for starch as before. Note the result.

4. In Exp. 21, carbon dioxide may be introduced into jar A by the burning of the taper. This would interfere with the lime-water test for carbon dioxide. Try to think of some way of getting over this difficulty.

CHAPTER XII

LEAVES

1. **Foliage leaves.** These are the ordinary green leaves of a plant. They usually consist of three parts, the **leaf-blade,**

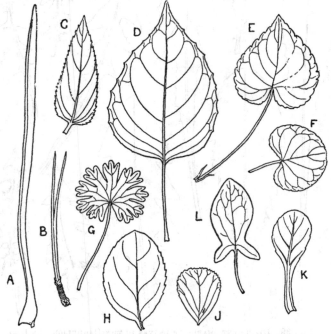

Fig. 38. SHAPES OF SIMPLE FOLIAGE LEAVES. A, Sea Plantain (linear). B, Pine (needle-shaped). C, Figwort (lanceolate). D, Enchanter's Nightshade (egg-shaped). E, Dog Violet (heart-shaped). F, Marsh Violet; G, Wild Geranium (both kidney-shaped). H, Brooklime (oval). J, Sun Spurge (reversed egg-shaped). K, Daisy (spatulate, i.e. shaped like a chemist's spatula). L, Sorrel Dock (spear-like). (Sept. ×5/8.)

the **leaf-stalk** or **petiole** and the **leaf-base**. Sometimes the petiole is absent, when the leaf is **sessile** (Lat. *sessilis*, sitting), e.g. Carnation, upper leaves of Groundsel (Fig. 4).

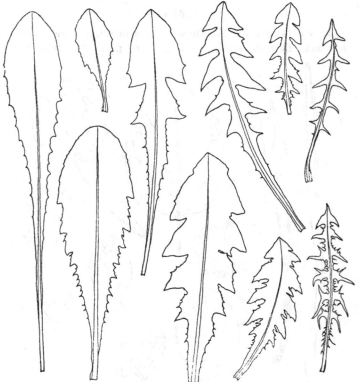

Fig. 39. DIFFERENT SHAPES OF LEAVES IN THE DANDELION. (× 1/2.)

Leaf-blade. The shape of the leaf-blade varies in different plants; some of the common shapes are shown in Fig. 38. When examining leaves you may find that you need special words or terms to describe what you see. A few of the terms used in describing the shapes of leaves are given under Fig. 38, for reference when you need them.

Each kind or species of plant usually has leaves of a particular shape; though the old saying that "no two blades of grass are exactly alike" is true for other plants as well as grasses. Sometimes the differences are striking; e.g. in the Groundsel and Harebell, the lower and upper leaves of the same plant are of quite different shapes (Figs. 3, 80). In other cases, e.g. Dandelion (Fig. 39), different plants of the same kind may vary greatly in leaf-form.

The edge or **margin** of the leaf-blade may be smooth or **entire** (Fig. 38 A), but is more commonly either **toothed** (often **serrate** or saw-like, Fig. 38 c) or **lobed** (Fig. 40 D, E). Some leaves are both toothed and lobed (Fig. 40 D); in Dandelion the margin shows great variation (Fig. 39).

The veins. Nearly all Dicotyledons have **net-veined** leaves (e.g. Sycamore, Fig. 7). The larger veins may be **pinnate,** with lateral branches arranged in a feather-like manner on either side of a central vein or midrib (e.g. most leaves in Fig. 38); or **palmate,** with several large veins spreading out from the top of the petiole, like the fingers of the palm of a hand (Figs. 38 G, 40 c).

In Monocotyledons the main veins are **parallel,** as in the Cock's-foot (Fig. 10) and other grasses. The network of veins so characteristic of Dicotyledons is rarely found: exceptions are the wild Arum (Lords and Ladies) and the Arrowhead.

The surface of the leaf-blade may be **smooth** (Laurel), **waxy** (Cabbage, Sea Holly), or **hairy.** In hairy leaves the hairs may be short or long, few or many and bristly or downy.

Simple and compound leaves. Most leaves are **simple,** that is, they have a single blade (Fig. 38). In **compound** leaves the leaf-blade is divided into several distinct **leaflets** each of which may look like a separate leaf (Fig. 40 A, B). You can easily distinguish between a compound leaf and a leafy branch because a leaf has a bud in its axil (Fig. 40 A), while a branch arises in

the axil of a leaf (Figs. 4, 34 B). A leaf may be slightly or deeply
lobed, but it is still regarded as a simple leaf so long as the spaces
between the lobes do not reach the midrib. The leaf of the Buttercup, for instance, is simple not compound (Fig. 40 c).

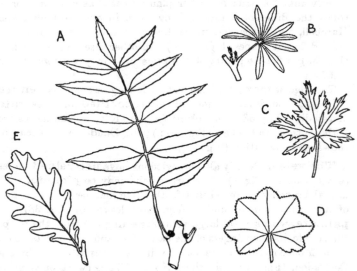

Fig. 40. Lobed and compound leaves. A, Ash. B, Lupin. C, Meadow
    Buttercup. D, Lady's Mantle. E, Oak. (Aug. ×1/3.)

Petiole. In most plants the petiole is rounded on the lower
surface, but flat or grooved on the upper; in **peltate** (shield-
shaped) leaves, such as the garden "Nasturtium," it is cylindrical.

Leaf-base. The leaf-base is usually broader than the petiole,
and serves not only to attach the leaf firmly to the stem, but
also to shelter the young axillary bud. **Sheathing leaf-bases**
are found in a number of plants, e.g. the Grass, Buttercup and
Parsnip families (Figs. 9–11, 41 A, B). In the Plane tree the leaf-
base fits over the young bud like a candle extinguisher.

Stipules. In many plants the leaf-base has a pair of out-growths called **stipules.** Some stipules are small, while others are large and leaf-like (Figs. 38 E, 41 E, 48), and take part in the work of photosynthesis.

The stipules and sheathing leaf-bases of many plants have the important duty of sheltering not only the axillary buds, but also the terminal bud with its delicate young leaves (Fig. 41).

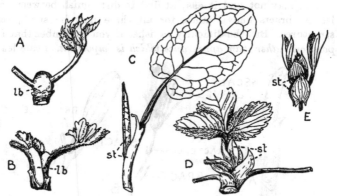

Fig. 41. SHEATHING LEAF-BASES. A, Gout Weed. B, Creeping Buttercup. *lb*, leaf-base. STIPULES. C, Dock. D, Strawberry. E, Meadow Pea. *st*, stipules. (April, × 5/8.)

2. Scale leaves. There are other kinds of leaves besides green foliage leaves. A **scale leaf** is a small, scaly structure, borne on a stem in the position of a leaf. It is generally white or brown in colour, and like an ordinary leaf may have a bud in its axil. Scale leaves are sometimes found on above-ground stems, but most often on underground stems, where they shelter both the terminal and axillary buds (Fig. 42 A). The scales of winter buds of trees are described on pp. 122–124.

3. Bracts and bracteoles. A flower usually arises as a bud in the axil of a leaf, which is called the **bract** of the flower. If there are leaves on the same stem as the flower itself, i.e.

between the flower and its bract, they are called **bracteoles.** As a rule, a Dicotyledonous flower has one bract and two bracteoles, and a Monocotyledonous flower one bract and only one bracteole (Fig. 42 B, C). Some flowers have no bracts at all, in which case we say that the bracts are undeveloped (Fig. 12). Bracts may be small and scale-like (Fig. 42 B), or large and leaf-like (Anemone, Fig. 54).

You may not find it easy at first to distinguish between scale leaves, bracts and stipules, for all three may be small, scaly structures. But the difficulty vanishes if you remember that *it is position rather than appearance which is important.* Scale leaves

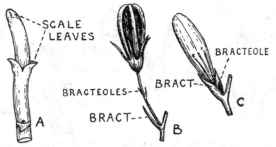

Fig. 42. SCALE LEAVES AND BRACTS. A, underground stem of Enchanter's Nightshade. B, Harebell. C, Montbretia. (Sept. × 1¼.)

have the position of ordinary leaves, bracts have flowers in their axils, and stipules are always two together, one on each side of the leaf-base.

4. **Floral leaves.** Sepals, petals and other parts of flowers are also leaves (i.e. **floral leaves**); these are dealt with in Chaps. XXII and XXIII.

We can now distinguish clearly (as we did in the case of the root, p. 43) between what a leaf is, and what a leaf does.

I. *Definition of what a leaf is.*

(*a*) A leaf is an external outgrowth from the side of a stem ;

that is, it arises on the outside of the stem (not inside as a root does, see p. 43), and is always lateral, not terminal in position.

(b) A leaf usually has a bud in its axil.

(c) Leaves are always developed on the stem in a definite order, the older ones below, the younger nearer the tip. The points on the stem at which leaves arise are called nodes.

These three things are true of almost all kinds of leaves: cotyledons, foliage leaves, scale leaves, bracts. Most leaves are also green and flat, but many scale leaves, bracts and floral leaves are not.

II. *What a leaf does.* A foliage leaf has two chief functions:

(a) Photosynthesis.

(b) Transpiration.

Other duties are often performed by leaves or parts of leaves, e.g. (c) young axillary buds are sheltered by leaf-bases, stipules or scale leaves; (d) the delicate growing tip of the stem is not provided with a protective cap such as we find in the root (p. 39), but is protected from drying up, and from other injuries, by the youngest leaves, which are tightly wrapped round it; (e) food may be stored in fleshy leaves, e.g. cotyledons of non-endospermic seeds (Chap. III), and bulbs (Chap. XVIII); the important functions of floral leaves are described later.

Unlike roots, leaves neither anchor the plant, nor (except in some water plants, Chap. XVI) do they absorb water and mineral salts.

HINTS FOR PRACTICAL WORK.

1. Collect a few leaves and, after examining them, describe them carefully. The following description will show you the proper way to describe a leaf. *Lady's Mantle* (Fig. 40 D); leaf simple, with long hairy petiole and stipules (not shown in figure); outline kidney-shaped; margin palmately lobed and finely serrate; tips of lobes blunt; larger veins palmate, smaller net-like; surface smooth except for a few hairs along margin.

2. Make a drawing of a leaf from the following description, and then see if you can identify it from your drawing; it is one of those figured in this

chapter. Leaf simple, with long petiole and no stipules; outline egg-shaped; margin with a few small teeth which point outwards; tip long and pointed; larger veins pinnate, smaller net-like; surface smooth.

3. Look for leaves of different shapes, either on the same plant or on different plants of the same kind, in the Lesser Celandine, Ivy, Dandelion, Hawthorn, Shepherd's Purse. Draw a few of them.

4. The veins of the leaf not only conduct water (p. 47) but also form a kind of framework or skeleton which strengthens the leaf. In winter or spring look for and examine "skeleton leaves" of Holly or Poplar, i.e. fallen leaves the soft parts of which have decayed, leaving only the veins behind.

5. Examine the leaves of Ash, Mountain Ash, False Acacia and Horse Chestnut, and notice that there are three reasons for saying they are compound leaves and not leafy branches :—(1) the petiole arises at a node and has a bud in its axil, (2) there are no buds in the axils of the leaflets, and (3) one of the leaflets is terminal in position, and cannot be a whole leaf, because leaves are always lateral.

6. Examine and draw the two kinds of scales found on a young branch of the self-clinging Virginian Creeper (cf. Fig. 50), and find out whether they are scale leaves, bracts or stipules. Look out for other examples of bracts, scale leaves, and stipules. Draw several kinds of stipules, e.g. Hawthorn, Bramble, Pansy, Vetch.

CHAPTER XIII

HOW FOLIAGE LEAVES GET LIGHT

CLIMBING PLANTS

How do the innumerable foliage leaves we see everywhere in nature get sufficient light for photosynthesis?

Form of leaves. In the first place, most leaf-blades are flat and thin, which allows light to penetrate the leaf easily. Further, most of the chlorophyll granules, which absorb the rays of light (p. 60), are near the upper surface of the leaf, where the light is strongest. This is why the upper surface is a darker green than the lower.

Arrangement of leaves on the stem. If you stand under-

neath a large tree, especially one with dense foliage, such as the
Sycamore or Horse Chestnut, you will see that most of the leaves
on the younger twigs are towards the well-lighted outside of the
tree. The same is true of many shrubs, e.g. a thick Privet or
Beech hedge. In the interior, where the light is dim, few leaves
are to be found.

Leaves are arranged on stems in various ways, but in most
cases the leaf-blades are placed so that they do not shade each
other very much. In some herbaceous plants the stem is very
short and the nodes crowded together, all the leaves arising from
about the level of the soil. Such leaves, e.g. Daisy (Fig. 81),
Dandelion (Fig. 39), and London Pride are said to be **radical.**
In these plants the leaves are spread out round the stem; the
broadest part of each leaf is towards the tip, and the lower leaves
are longer than the upper. All these features help the leaves to
obtain light.

Most plants, however, have more or less elongated internodes,
so that the leaves are not crowded together. Sometimes the
leaves are **opposite,** i.e. one pair at each node, as in the Sycamore and Privet. Opposite leaves are usually arranged on the
stem in four vertical rows, each pair of leaves being at right
angles to the pairs immediately above and below (Figs. 6, 43 A).
The majority of plants, however, have **alternate** leaves (i.e. only
one at each node), arranged in a spiral round the stem. Simple
examples of the alternate arrangement are the Elm and grasses
(Figs. 44, 9), where half the leaves are on one side of the stem
and half on the other, i.e. in two vertical rows. In other cases
the arrangement is more complicated, the leaves being in three,
five, eight or even more vertical rows, five rows being very
common. As a rule the strongest light comes from above, so
this spreading out of the leaves on different sides of the stem
to a great extent prevents the shading of one leaf by another
(Figs. 2, 81). In the Sycamore and other plants, the lower leaves
on a twig have longer petioles than the upper; this too helps to
prevent shading of the lower leaves.

Movements of leaves. Compare an upright twig taken from the top of a Privet hedge with a horizontal one from the side of the hedge (Fig. 43). In both twigs the leaves are joined to the stem in four rows. This is easily seen in the upright twig, but in the other the short petioles have twisted so as to bring the leaf-blades into two rows, with their upper surfaces facing the sky. In both cases the leaves have arranged themselves in the best position to receive the light coming from above. Most young leaves have a similar power of adjusting their positions so as to face the direction from which light is coming (p. 86).

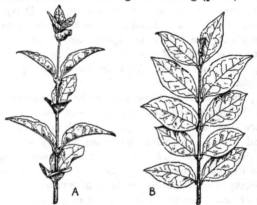

Fig. 43. PRIVET TWIGS. A, from top of hedge; seen from the side. B, from side of hedge; seen from above. (Aug. × 1/4.)

Leaf mosaics. We see then, that the thin, flat form of leaves, their arrangement on the stem, the varying lengths of their petioles, and their power of twisting or turning so as to face the light, all help leaves to get light. In many cases the result is that a more or less continuous pattern is formed, into which the different leaf-blades fit, rather like the pieces of a Roman mosaic pavement. Such patterns formed by leaves are known as **leaf mosaics** (Figs. 2, 44, 81).

Functions of stems.   In most plants the stems have two chief duties :

(1) To carry water and other substances from the roots to the leaves, or from the leaves to other parts of the plant.

Fig. 44.   LEAF MOSAIC OF ELM.   (Aug.)

(2) To support the leaves and flowers, and to spread them out so that they can perform their functions properly.   In order to support the weight of the leaves, especially when wind is blowing, most stems require to be tough and strong.

Competition.   In nature, plants are often much crowded together, and compete with each other for light and other necessaries of life.   Trees and other tall plants obtain light by growing to a greater height than their neighbours.   In order to do this, and to support and spread out their numerous leaves, they must spend a great deal of material in building large, strong trunks and branches. Some plants, however, obtain the same advantages without forming massive stems. These are the

Climbing plants,

which support themselves by clinging to some external support such as other plants, rocks or walls. Mr Charles Darwin found that there are four classes of climbing plants.

1. **Twining plants,** in which the stem twines spirally round a support. The latter is usually the stem of another plant, though any support will do, provided that it is upright, and neither too thick nor too smooth. The tip of the twining stem does not grow straight upwards, but traces a spiral path in

Fig. 45. GREAT BIND-
WEED TWINING ROUND
REED GRASS. (July,
× 1/3.)

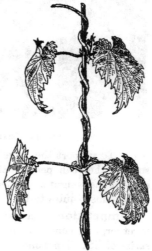

Fig. 46. HOP TWINING ROUND
STEM OF PRIVET. (Sept. × 1/4.)

the air, in this way gradually coiling round and round the support (Fig. 45). Only the young, growing internodes have the power of twining; as they become older, the movement ceases. In the Black Bryony, Hop and Honeysuckle the tip of the stem moves round with the sun, i.e. from E. to S., S. to W., W. to N., then back to E. and so on. This is the same direction as that taken by the hands of a watch (Fig. 46). In the Scarlet Runner Bean and Great Bindweed (Fig. 45) the stem moves against the sun, i.e. in the opposite direction to the hands of a watch.

2. **Tendril climbers** have special clinging organs which attach themselves to any suitable support. These organs are

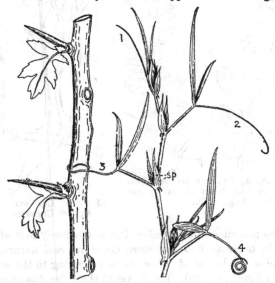

Fig. 47. MEADOW PEA CLINGING TO HAWTHORN. 1–4, tendrils of different ages; *sp*, stipules. (Aug. × 1.)

long thread-like structures called **tendrils**. In the Meadow Pea (Fig. 47) the leaf is compound, with two stipules, two lateral leaflets and a terminal tendril. At first the tendril is nearly straight, but if it comes into contact with a rough object such as a twig, the tip bends, and gradually curls round the support (Fig. 47, 1–3). The tendril now becomes hard and woody, and cannot easily be uncurled. By attaching itself to other plants in this way, even a slender, delicate plant like the Pea may climb to the top of a hedge, where there is plenty of light.

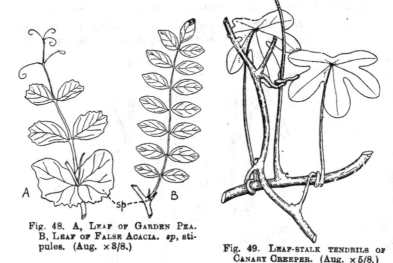

Fig. 48. A, LEAF OF GARDEN PEA. B, LEAF OF FALSE ACACIA. *sp*, stipules. (Aug. × 3/8.)

Fig. 49. LEAF-STALK TENDRILS OF CANARY CREEPER. (Aug. × 5/8.)

The tendril of the Meadow Pea is obviously part of the leaf; but in order to find out more about its real nature, we must examine the leaves of other plants belonging to the same family (Bean family, p. 188). The leaves of the False Acacia and Garden Pea (Fig. 48) are both constructed on the same general plan as that of the Meadow Pea. All three leaves are compound and all have stipules. Yet in some ways they are very different. The two Peas have leaf-like stipules, while those of the False Acacia

are spiny. In the latter all the leaflets are alike, but in the
Garden Pea the places of the terminal leaflet and two or more
pairs of lateral leaflets have been taken by tendrils (Fig. 48).
We may conclude from this comparison that the tendrils in the

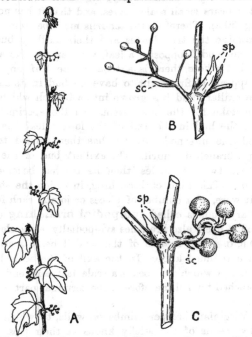

Fig. 50. A, Self-clinging Virginian Creeper. B,
Young branched Tendril. C, Older ditto. *sp,*
stipules; *sc,* scale leaves. (July, A × 1/3; B, C × 2½.)

Garden Pea are modified terminal and lateral leaflets, and in
the Meadow Pea terminal leaflets only.

All tendrils, however, are not modified leaflets. In Clematis,
the climbing Nasturtium and the Canary Creeper (Fig. 49) the
leaf-stalk forms a kind of tendril. The branched tendrils of the

self-clinging Virginian Creeper (Fig. 50) are more complicated.
They arise opposite to, and not in the axils of leaves, so at first
sight it might be supposed that they themselves were modified
leaves. But if you examine a young tendril carefully, you will
see that it bears small scale leaves, and that it has no bud in its
axil (Fig. 50 B). Therefore the tendrils must be stems, not leaves.
Now examine the true leaves and their axillary buds. Where
there is no tendril opposite a leaf, you can see the axillary bud
plainly, but where there is a tendril, the leaf opposite to the
tendril appears at first sight to have no bud in its axil. In this
case the axillary bud has grown into a branch which you might
easily mistake for the main stem. In the specimen shown in
Fig. 50 A the bud in the axil of the lowest leaf has grown and
produced one internode, its tip has then turned to one side,
forming a branched tendril. The axillary bud of the second leaf
has formed two internodes, then its tip has become a tendril,
and so on. This kind of branching, in which the stem is made
up, as it were, of a number of pieces or joints, each formed from
an axillary bud, is called **sympodial branching** (cf. p. 119).
The tendril itself also branches sympodially, as is clearly seen in
Fig. 50 B, C. Each branch of the tendril bears a scale leaf and
ends in a rounded knob. In the axil of the scale leaf another
branch arises which also bears a scale leaf and ends in a knob.
The branched tendril therefore is the terminal part of an axillary
branch.

The Virginian Creeper climbs on walls, to which its tendrils
cling by means of the sticky knobs at their tips. The knobs
are sensitive to contact, and if they press against a solid object,
they grow into flattened discs which adhere firmly to the object
(Fig. 50 C).

3. **Hook scramblers.** The third class of climbers includes
plants like Cleavers (= Goose-grass), Roses and Brambles (Fig. 51),
which are often found in hedges. The stems and sometimes the
leaves of these plants are furnished with recurved prickles (p. 54),

which allow the shoots to grow forward amongst other plants, but prevent them from slipping back. By hooking on to other plants in this way, the shoots can scramble to positions where their leaves can obtain light. These plants are not attached as firmly as twiners or tendril climbers; in fact they scramble rather than climb, and so may be called "**hook scramblers.**"

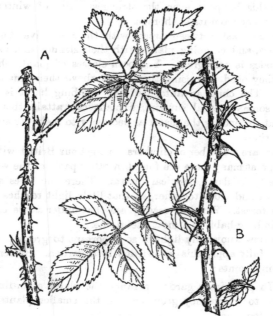

Fig. 51. PRICKLES OF BRAMBLE (A) AND ROSE (B).
(Aug. × 1/2.)

4. **Root climbers.** The Ivy is the largest of our native British climbing plants, and the only one belonging to the fourth class, i.e. those which climb by roots. It is found on rocks, walls or tree trunks, often climbing to the tops of tall trees. Numerous small adventitious roots are formed on the stem of the Ivy, on

the side next the support. These roots grow into crevices in the bark or wall, and firmly anchor themselves there. In the case of Ivy growing on a tree trunk, the older stems often become anchored in another way as well. Where the branches cross one another, they fuse together, forming an exceedingly strong network of stems, which in many cases clasps the entire trunk. When this happens, even the strongest gales of winter cannot tear the Ivy from its anchorage.

The contrast between trees, with their massive trunks and branches, and climbing plants with their slender stems and long internodes is very great. Yet both classes of plants obtain the light they need by raising themselves above their smaller neighbours. The chief advantage of the climbing habit is that, by using some external support, a plant can attain this position without expending the material necessary to make its stems self-supporting.

There are a number of climbers amongst our British wild plants, but a great many more are found in other parts of the world, and especially in damp, tropical forests. There the trees are often very tall and close together, so that little light reaches the floor of the forest. You will readily see that under these conditions the climbing habit is a most useful one.

We now know why light is so important to green plants, and can usefully apply this knowledge in a garden. Remember the following points:

i. In planning a garden, arrange the plants according to the heights to which they grow, or else the smaller plants may be shaded too much.

ii. Don't overcrowd the plants. Thin out where necessary, so as to give each plant plenty of light and room to grow.

iii. All plants need light, but some need more than others. Peas and Potatoes, for instance, must have plenty of light, but Ferns, Mint, Spinach etc., may be grown in the shadier parts of the garden.

HINTS FOR PRACTICAL WORK.

1. Examine some dense shrubs and hedges in summer. Notice the dim light and the absence of leaves in the interior. If you have a photographer's actinometer, see how many seconds it takes to match the standard tint, first on the outside of the hedge and then in the interior. This will tell you how much brighter the light is on the outside.

2. Examine and draw examples of leaf mosaics, e.g. (a) Plantain, Daisy or London Pride, (b) Groundsel, and (c) an upright and a horizontal twig of Sycamore (cf. with Fig. 43). In each case make a list of features which help the leaves to get light.

3. Examine and draw as many examples of climbing plants as you can find. It will be enough to draw a small piece of each, showing the exact form and position of the climbing organ.

4. Grow a few Peas and Scarlet Runner Beans in pots, and provide the seedlings with supports. Observe and make notes on the way in which the tendrils or stems coil round the support. How long does it take for them to do this?

CHAPTER XIV

THE WAYS IN WHICH A PLANT USES ITS FOOD

GROWTH AND RESPIRATION

Growth. In Chaps. IV and V we learned three important facts about the behaviour of seedlings: (1) *that they grow,* (2) *that they respire,* and (3) *that they do work,* such as forcing apart the particles of soil when the radicle and plumule come out of the seed. We found too that food is stored in all seeds, and that this food is used during the growth of the young plant (Exp. 1, p. 29). Before the stored food is completely used up the seedling has become an independent plant, and can make more food for itself, which is gradually built up into the living substance of the plant. Food then provides material for the growth of the bodies of plants and animals.

Respiration. Growth, however, is not the only purpose for which food is necessary. We might perhaps infer this from the fact that in the case of human beings even old people need food, though they have stopped growing. When we have gone for rather a long time without food, we begin to feel tired, but when we have satisfied our hunger we feel more energetic and ready to work again. We must try to find out whether food has any effect of a similar kind on plants.

EXP. 22. Exp. 4 (p. 83) showed us that germinating seeds respire like ourselves, and that during respiration oxygen is used up and carbon dioxide set free. Repeat Exp. 4, using parts of older plants, e.g. leaves, opening flower buds, or even a bunch of roots. In these cases too oxygen is used up and carbon dioxide given off. It has been found that water also is given off, but this is not so easy to prove. N.B. If green parts of plants are used in this experiment, the jars must be kept in the dark.

Not only germinating seeds, but all living parts of plants and animals respire. We know that an animal deprived of air is suffocated and dies. This is because in the absence of air, or rather the oxygen of the air, respiration stops. For the same reason, in Exp. 2 (p. 30), if the seeds in jar 2 were kept without air for a considerable time, not only would they not grow or germinate, but they would actually die. *Respiration in fact is necessary, not only for growth, but also for life itself.*

EXP. 23. Take a quantity of germinating Barley grains or Pea seeds, and pack them round a thermometer in a thermos flask (or Dewar vacuum flask), so that the bulb of the thermometer is in the middle of the seeds. Put a plug of cotton-wool in the neck of the flask. As a control experiment set up a similar apparatus with seeds which have been killed by boiling in water containing a little antiseptic, such as permanganate of potash. It will soon be noticed that the temperature of the actively respiring seeds is higher than that of the dead ones. Energy, in the form of heat, is set free during the respiration of germinating seeds, and it is this energy which raises the temperature of the thermometer.

It has further been found by experiment that if seeds germinate and respire in the dark, the seedlings after a time contain less solid material than the seeds from which they came. Evidently some of the material has disappeared.

We see then that several things happen during the respiration of seedlings, older plants, and animals:

(1) Oxygen is taken in, and carbon dioxide and water given off (Exps. 4, 22).

(2) There is a loss, or using up, of solid material.

(3) Energy (in the form of heat) is set free (Exp. 23).

Energy. Heat is one form, light and electricity are other forms of energy. It is not at all easy to explain what is meant by energy, except by saying that it is concerned with the doing of work. In fact, the words "energy" and "energetic" come from the same word as the Greek verb *energeo*, I work.

Why respiration is necessary. We shall most easily understand why living beings need to respire if we compare a plant or animal with a machine such as a railway engine or an aeroplane. When such a machine is moving, it is doing work, and therefore spending energy. This energy is obtained from the burning or combustion of fuel which contains carbon—coal in one case and petrol in the other. When coal (or petrol) burns, several things happen:

(1) Oxygen is taken in (coal will not burn without oxygen), and carbon dioxide and water given off.

(2) The coal is gradually used up, and needs to be replenished.

(3) The energy contained in the coal is set free (partly as heat and light), and can be used to do the work of driving the engine. The faster the engine travels, or the heavier its load, the more energy is required, and the more coal will be used.

All this is very much like what happens during respiration. A plant or animal is really a living machine which, if it is to do work, must be supplied with energy. This energy is obtained from carbon-containing food such as sugar or starch. The food is oxidised during respiration just as coal is oxidised when it burns, in both cases energy being set free. You have noticed that the harder you work, for instance in running, the more quickly you breathe. Rapid breathing means that more oxygen is taken in, more food oxidised or respired in the body, and therefore more energy set free.

Respiration and food. The plant then uses the food it has made for two purposes, growth and respiration. The food is not always used immediately, but may be stored for future use (p. 111). We can now say exactly what we mean by the terms respiration and food.

Respiration *is really a kind of slow combustion or oxidation of carbon-containing food in the body, which results in the setting free of energy stored in the food.*

A food *is any substance which can be used directly by a living organism, either* (a) *for growth, or* (b) *as a source of energy which can be used by the organism in doing work.*

Carbon dioxide, water and salts are not really foods, but only the raw materials out of which true foods such as starch, sugar, oil etc. can be made by the plant.

Work done by plants. Plants work silently, but none the less are they constantly doing work of various kinds, for all of which energy is required. Roots bore their way through the soil; stems, as they grow, lift themselves into the air; food and water have to be carried from one part of the plant to another; and finally, many plant organs carry out different kinds of movements—these we shall consider in the next chapter.

Photosynthesis and respiration are two of the most important functions of green plants. Without photosynthesis the plant could not make food, and without respiration food would be of little use to it. In many ways these two processes are almost the exact opposites of one another. In photosynthesis food is built up from carbon dioxide and water, oxygen being set free. In respiration oxygen is taken in and food oxidised and broken down, carbon dioxide and water being set free. During photosynthesis, energy is absorbed from sunlight by chlorophyll, and stored in the food which is made. This energy is set free again during respiration, and part of it used in doing work of various kinds.

Photosynthesis takes place only in green plants or parts of plants, and only in light. Respiration occurs in all living parts of plants (not only in green parts) and animals, and in both light and darkness. At night a plant respires, taking in oxygen and giving off carbon dioxide. During the day respiration and photosynthesis are going on at the same time, carbon dioxide being taken from the air and oxygen given off. This result may seem surprising, for it is just what would happen if photosynthesis alone were going on. The explanation is that in bright light photosynthesis is very much more rapid than respiration. In bright light the plant not only uses all the carbon dioxide set free by respiration, but takes in more from the air as well.

HINTS FOR PRACTICAL WORK.

1. In Exp. 22 be sure to keep any jars containing green parts of plants in the dark. Why is this precaution necessary? If you have read this chapter carefully you can easily answer this question. Gas jars with well greased glass plates may be used in this experiment, as in Exp. 21.

2. Young Dandelion flower heads may be used in Exp. 23 instead of seeds. Sunlight must not be allowed to fall on the apparatus. Read the thermometers at intervals of 15 minutes.

3. Draw and describe the apparatus used in each experiment, and keep full records of the results.

CHAPTER XV

THE MOVEMENTS OF PLANTS

Unlike animals, plants do not move about actively from place to place. They cannot, for instance, move bodily into the shade if the sun is shining fiercely on them. Yet none the less movements of many kinds are performed by the organs of plants. As a rule plant movements are slower than those of animals, but even slow movements involve the doing of work, and therefore the spending of energy.

If we touch one of the compound leaves of the Sensitive Plant (often grown in hothouses), the little leaflets fold together and the whole leaf rapidly droops. We say that the leaf "responds to the stimulus" of the touch by moving into a fresh position. A **stimulus** is any agency, usually outside the plant, which brings about some **response** or behaviour in the plant. A whip applied to a horse acts as a stimulus, to which the animal responds by moving more rapidly. Practically all the plant movements we are going to consider are responses to stimuli of various kinds. We can speak of the tendril of a climbing plant responding to the stimulus of contact with a rough object, by curling round the object (Chap. XIII).

"Sleep movements" of leaves. The leaves of many plants have different positions by day and night. The leaves of

Fig. 52. DAY AND NIGHT POSITIONS OF CLOVER LEAVES.
(Aug. × 1.)

Clover, for instance, are flat and expanded during the daytime, but towards night the two side leaflets fold together and are roofed over by the third leaflet (Fig. 52). The movements which bring such leaves into the night position are called **"sleep movements,"** though the leaves do not really go to sleep in the sense that we ourselves do. The expanded day position is the best for photosynthesis; the folded "sleep" position probably keeps the leaf warmer at night.

EXP. 24. Darken a clover plant during the daytime by covering it with a large flower pot or a box. Plug with plasticine any holes or cracks in the pot and scatter soil round its bottom edge, to keep out all light. After an hour or two the leaves will be found in the night position; this shows that the stimulus inducing the movement is the change from light to darkness.

Daily movements of flowers. Many flowers open by day, especially in bright, sunny weather, and close at night or in dull, wet weather, e.g. Crocus (Fig. 53). Some flowers not only close at

Fig. 53. MOVEMENTS OF FLOWERS OF CROCUS. (Mar. × 2/3.)

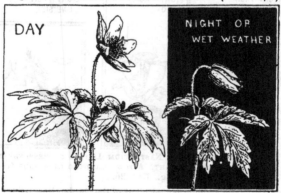

Fig. 54. MOVEMENTS OF FLOWERS OF WOOD ANEMONE. (Mar. × 2/3.)

night, but also hang downwards, e.g. Wood Anemone (Fig. 54). As a rule these daily movements result in the flowers being open when honey-seeking insects are most active, and closed when the pollen might be liable to injury from dew or rain (cf. Chaps. XXIV, XXV)

Growth movements. We have seen that during growth roots, stems and leaves place themselves in certain definite positions (Chaps. IV, VII, XIII). If we think about it, this seems a wonderful thing. What is it that causes the radicles of all germinating seeds to grow downwards, and the plumules upwards, no matter in what positions the seeds were sown (cf. Fig. 21)? Is it possible that the organs grow in these directions in response to influences or stimuli acting on the plant from outside? We must study the behaviour of plants a little further, and find out if these questions too can be answered by the method of experiment.

Effect of light on stems and leaves.

EXP. 25. Grow some Broad Beans out of doors, and others indoors by a window. The stems of those in the open, where light comes chiefly from above, are upright. In the window plants, however, which are lighted only from one side, the stems bend and grow towards the light (Fig. 56 C).

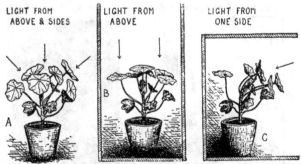

LIGHT FROM ABOVE & SIDES   LIGHT FROM ABOVE   LIGHT FROM ONE SIDE

Fig. 55. MOVEMENTS OF NASTURTIUM LEAVES IN RESPONSE TO THE STIMULUS OF LIGHT. Arrows show the directions from which light is coming. EXP. 26.

EXP. 26. Keep a young potted plant of the Garden Nasturtium for a day or two in a well-lighted spot in the open. Notice that each leaf-blade takes up a position at right angles to the strongest light it can get (Fig. 55 A). Now vary the experiment by using boxes blackened on the inside, so that the plant is exposed, first to light coming only from above and afterwards to light from one side. In both cases the petioles curve and place the leaf-blades again at right angles to the light (Fig. 55 B, C).

From these two experiments we learn that light does influence the positions taken up by stems and leaves. In most cases (the exceptions need not concern us now) stems respond to the stimulus of light by growing towards it, while leaf-blades arrange themselves at right angles to the strongest light.

Effect of gravity on stems and roots. The growth curvatures or movements of plants, however, cannot be caused by light alone, for even in complete darkness, a germinating seed never makes a mistake; its plumule always grows upwards and its radicle downwards. Seedlings and older plants too seem able, in some wonderful way, to appreciate the difference between the up and the down directions.

EXP. 27. Grow a Broad Bean plant in a pot till it is about five inches high; then lay the pot on its side in a dark cupboard. At first the stem is

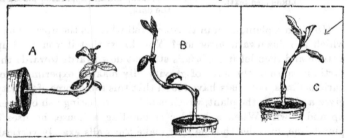

Fig. 56. GROWTH MOVEMENTS OF STEMS OF BROAD BEAN IN RESPONSE TO THE STIMULI OF GRAVITY (A, B) AND LIGHT (C). EXPS. 27, 25.

horizontal, but after a few hours it will be found to have curved so that the younger parts are once more in an upright position (Fig. 56 A). After the plant has grown for a day or two in its new position place the pot vertically again, and leave in the dark as before; a new bend appears (Fig. 56 B). In whatever position the plant is placed, the growing part of the stem curves or moves till it is upright once more. If you tried this experiment with other plants, e.g. a Geranium or a pot of Mustard seedlings, the result would be the same.

EXP. 28. Select three or four young, healthy Broad Bean seedlings, and place them with their plumules and radicles horizontal. In a short time the plumules and radicles will be found to have curved into a vertical position; but while the plumules have curved upwards like the older stem in Exp. 27, the radicles have curved until they point straight downwards.

During this experiment the seedlings must be kept moist and in the dark. A suitable damp chamber for this and other experiments (e.g. Exp. 5) is

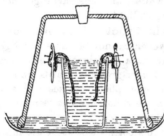

Fig. 57. DAMP CHAMBER FOR SEEDLINGS (see text).

shown in section in Fig. 57. Pieces of cork are fitted on the edge of a small tumbler of water, and the seeds pinned to the corks in the position required. The seedlings are supplied with water by loose cotton wicks arranged as in Fig. 57. The air is kept moist by standing the tumbler in a shallow dish or plate of water, and covering it with an inverted flower pot. The porous pot sucks up water which evaporates from its sides. The hole in the pot is corked, and the apparatus placed in the dark.

How can a plant, even in the dark, tell which is the upward and which the downward direction? You know that if you hold up a stone and then let it go, it falls straight downwards towards the earth, because of the force of gravity. By making experiments of various kinds, botanists have found that this same force of gravity gives a signal to the plant, which enables it to distinguish between up and down. When a builder is building a house, he uses a plumb-line, in order that he may make the walls exactly vertical. Gravity acts on the leaden weight of the plumb-line, and so tells the builder which is the vertical direction. A plant uses gravity in much the same way as a builder, i.e. to find out which direction is up and which is down. You must not make the mistake of supposing that it is the mere weight of the root which causes it, when placed horizontally, to bend in the direction that a stone falls, i.e. towards the earth. If this were the case, stems would

grow downwards too. In reality, gravity acts merely as a stimulus
or signal to the plant, telling it which is up and which is down.
The machinery of the plant responds to the stimulus in such a way
that the stem grows upwards, the main root grows downwards,
and the lateral roots spread outwards. *The various organs of a
plant, then, may respond in different ways to the same stimulus.*
We have seen another example of this in the fact that stems and
leaves respond in different ways to the stimulus of light.

Effect of moisture on roots.

EXP. 29. Remove the bottom from a shallow wooden box, and replace
it by string netting. Fill the box with damp moss and plant a number of
Pea seeds in it. Hang up the box with one side higher than the other in a
dark cupboard, and keep the moss damp. As the seeds germinate, their
roots grow downwards in response to gravity, till they reach the bottom of
the moss. Now, however, instead of continuing to grow downwards into the
dry air, as they would if the air were very moist (cf. Fig. 57), they curve
back towards the damp moss, and may even re-enter it. Evidently the effect
of moisture may be even greater than that of gravity. In nature roots tend
to grow towards the damper parts of the soil, where plenty of water may
be obtained.

The remarkable growth movements we have just studied are
evidently of use to the plant. The responses of roots to the stimuli
of gravity and moisture guide
them into the soil, from which
they can absorb water and
salts. On the other hand, the
stimuli of gravity and light
guide the stems and leaves
away from the dark soil, and
towards the light necessary
for photosynthesis.

Fig. 58. Positions of Flower bud (A),
open Flower (B), and Fruit (C) of
Bluebell. (A, B, May; C, July, × 1.)

**Other growth move-
ments.** Plant organs often
take up different positions as they become older, or reach a
certain stage of development. For instance, in the Bluebell the

flower buds are upright at first, but curve downwards as they open ; finally, by the growth of the flower-stalk, the developing fruits are brought again into the erect position (Fig. 58). Again,

Fig. 59. MOVEMENTS OF FLOWERS OF WHITE CLOVER.

Fig. 60. BENDING OF DAFFODIL FLOWERS TOWARDS LIGHT.
Arrow shows direction of light.

the flowers of the White Clover move downwards, one by one, after pollination (Fig. 59. Cf. Chap. XXV).

If you look at a flower border by the side of a house, you will notice that most of the flowers face away from the house and towards the light. Daffodil flowers for example are erect in the bud, but bend towards the light before opening (Fig. 60). This turning towards light makes flowers more easily seen by insects (Chap. XXV).

The position of young leaves too is often different from that of older ones. In many cases unfolding leaves have a more or less vertical position, either erect or drooping (Figs. 8, 61, 99, 100). Later on they move into the best lighted positions (Chap. XIII).

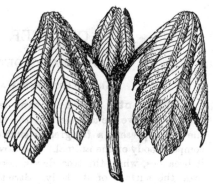

Fig. 61. YOUNG LEAVES OF HORSE CHESTNUT IN THE "HALF-OPENED UMBRELLA" POSITION. (April, × 3/8.)

Summary. 1. Plant organs are sensitive to stimuli of various kinds, e.g. gravity, light, moisture and contact (e.g. tendrils).

2. The organs usually respond to these stimuli by moving into positions in which they can best perform their respective functions.

3. All these movements are examples of work done by plants. Respiration provides the energy needed for this and other kinds of work.

HINTS FOR PRACTICAL WORK.

1. Sow the seeds for Experiments 25 to 28 early enough, so that the seedlings will be ready when wanted.

2. As before, keep a full record of all experiments performed, and the results. It will be interesting to make notes of the times required for the different movements to take place.

3. Notice day and night positions of leaves of Clover, Wood Sorrel, Scarlet Runner Bean, False Acacia, seedlings of Cress, and other plants.

4. Examine a number of flowers, such as Daisy, Eschscholzia, Tulip, Pimpernel, White Campion, Clematis, and see if they close at night or not. Draw one or two of them in day and night positions.

5. Compare the ways the flowers face in (*a*) an open, well-lighted flower bed, and (*b*) a border with a house or hedge behind it.

6. Notice the positions of the leaf-blades of Ivy and Virginian Creeper growing on a wall, and of plants growing under trees (e.g. in a wood), etc. and see how they are arranged with respect to light.

CHAPTER XVI

HOW PLANTS OBTAIN AIR

LAND PLANTS AND WATER PLANTS

Land plants. If we compare, let us say, a Sycamore tree with a Cow, we see at once that the many branches and leaves of the plant present a far greater surface to the air than does the compact body of the animal. The Cow draws air into its lungs as it breathes, while the tree slowly takes in the gases it needs, all over the surface of its body. Strictly speaking, plants do not "breathe." Breathing is the taking in and giving off of air from the lungs; it is not the same as respiration. Plants respire, but animals breathe and respire. Leaves and young stems take in air through the **stomata,** and the older stems of trees through the **lenticels,** i.e. porous places in the bark which appear as small oval spots on the younger twigs (Figs. 88, 91). Roots have neither stomata nor lenticels, but obtain oxygen dissolved in the water which they absorb. That roots need oxygen for respiration may be inferred from the fact that an ordinary potted plant is gradually suffocated if we over water it, and prevent its roots from getting air. Fig. 62 shows a number of dead Birch trees standing in a lake which had been dammed up to serve as a reservoir. As the level of the water rose, the roots of the Birch trees growing on its banks were deprived of air, in consequence of which the trees gradually died.

Many plants, however, flourish by the sides of lakes and rivers, with their roots always covered by water, while others may live in the water itself. These plants need oxygen for respiration and carbon dioxide for photosynthesis, just like land plants. We must find out how they obtain these gases.

Fig. 62. BIRCH TREES KILLED BY RAISING WATER LEVEL. (Sweden, Aug.)

Water plants. True **water** or **aquatic plants** live in the water, while **marsh plants** grow along its margins or in any waterlogged soil. We will start with a few examples of water plants.

Most of the Pondweeds live entirely submerged in ponds or rivers (Fig. 63 A). They have slender stems and extremely thin, delicate leaves. Take one out of the water and notice that it very soon withers. This is because the skin has no waterproof covering such as a land plant has. If we cut the stem across and look at the cut end with a lens we see that it contains many large air spaces (Fig. 69 A).

The submerged leaves of the Water Crowfoot are divided into numerous fine, thread-like portions. Some kinds of Water Crowfoot have thicker, more or less rounded leaves as well, which float on the surface (Fig. 64 A). In this case too the stems are provided with large air spaces.

All the foliage leaves of the Water Milfoil are submerged and finely divided (Fig. 64 B).

The Shoreweed (*Littorella*) and Water Lobelia (Fig. 65), which grow in shallow water, often in mountain lakes, are examples of another type of water plant. They have short stems bearing a crown of stiff, fleshy leaves, the Shoreweed often producing creeping branches from which new buds arise (Fig. 65 A). In these two plants the leaves themselves contain large air spaces (Fig. 69 D).

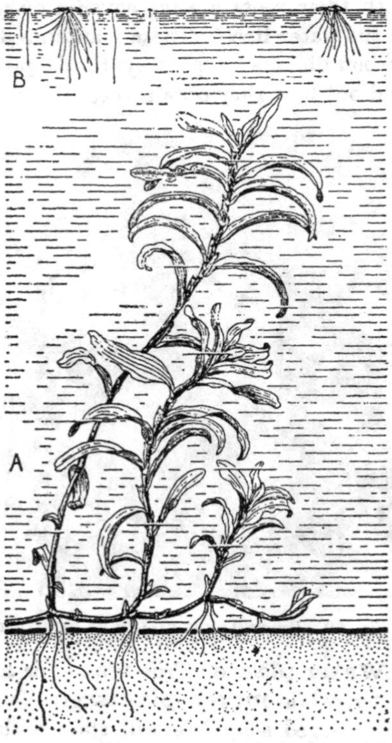

Fig. 63. A, PONDWEED WITH ROOTS IN MUD. B, FLOATING DUCKWEED. (Aug. × about 1/3.)

The White and Yellow Water Lilies have fleshy, creeping stems **or rhizomes** (p. 111) rooted at the bottom of the water, and

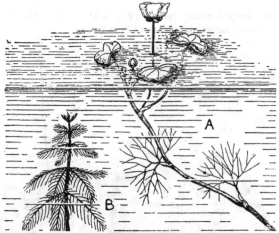

Fig. 64. A, Water Crowfoot with submerged and floating leaves. B, Water Milfoil. (July, × 1/2.)

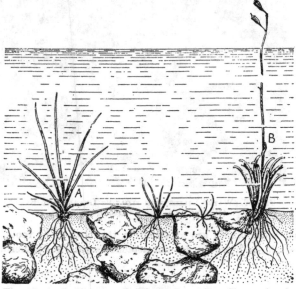

Fig. 65. A, Shoreweed. B, Water Lobelia in Fruit; break
in stem indicates that the stem is longer than shown in
the drawing. (Sept. × 1/3.)

large, leathery floating leaves (Fig. 66). The roots and petioles contain many large air spaces (Fig. 69 B, C)

Fig. 66. WATER LILIES WITH FLOATING LEAVES, AND REEDS.
(Wicken Fen, Cambridgeshire. July.)

All these plants are attached to the bottom by roots. Some water plants, however, are not attached at all, but either float on the surface of the water, e.g. Duckweed (Fig. 63 B), or swim below it, e.g. the Bladderwort (Fig. 67).

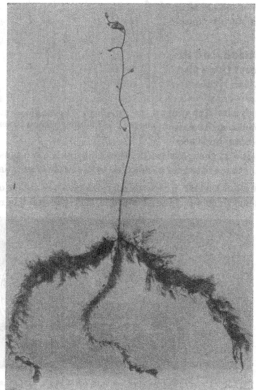

Fig. 67. COMMON BLADDERWORT WITH FINELY DIVIDED, SUBMERGED LEAVES, AND FLOWERS ABOVE THE WATER. (Wicken Fen. July, × about 3/8.)

The common Arrowhead found in ponds and streams is another interesting water plant. It has three kinds of leaves, thin ribbon-shaped submerged leaves, thicker oval floating leaves, and leaves shaped like an arrowhead which come above the water. Leaves intermediate in shape between these three main types can often be found (Fig. 68).

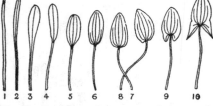

Conditions of life in water. If we plant either a land plant in the water, or a water plant on dry land, it will die. Yet both alike are green plants and need

Fig. 68. LEAVES OF ARROWHEAD. 1–4 are submerged; 5–8 floating; 9 and 10 aerial. (July, × 1/6.)

water, salts, oxygen, carbon dioxide and light. We must see how water plants are fitted for or adapted to life in their peculiar habitat.

Water and salts. If a submerged plant is taken out of the water it soon dries up, for its skin is very thin and is not water-

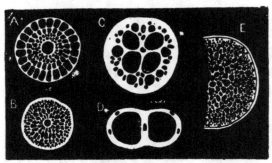

Fig. 69. CROSS SECTIONS OF ORGANS OF WATER AND MARSH PLANTS, ON BLACK BACKGROUND TO SHOW AIR SPACES. Veins are shaded. A, stem of Pondweed. B, root, and C, petiole of White Water Lily. D, leaf of Water Lobelia. E, stem of Flowering Rush. (Sept. × 5.)

proof. But in the water the thin skin is very useful, for it allows water and salts to be absorbed over the whole surface of the plant.

Air supply. The chief difficulty of a water plant is to obtain enough air, and especially oxygen. Water can dissolve a good deal of carbon dioxide, but very little oxygen, yet submerged plants must obtain these necessary gases from the water in which they live. The submerged leaves of water plants are usually either very thin or else finely divided (Figs. 63, 64), so that every part of the leaf is continually bathed with water. This makes it easier for them to absorb gases dissolved in the water. Another peculiarity of water plants is that their stems, roots and in some cases leaves contain many air spaces, far larger than those of land plants (Fig. 69). These spaces are very important, for oxygen given off during photosynthesis is stored in them, and can be used later for respiration. Another function of the air spaces is to carry oxygen down to the roots or stems in the mud, where there is very little oxygen indeed. Floating and aerial leaves take in oxygen directly from the air through their stomata (submerged leaves have no stomata), and this oxygen can travel downwards through the air spaces.

Light. The floating and aerial leaves obtain both light and air as easily as the leaves of land plants. Water, however, absorbs so much light that plants a few feet below the surface are practically living in the shade. The deeper the water the less light there is, and the fewer the plants which grow near the bottom.

Buoyancy. The stems and petioles of aquatics are usually weaker and more delicate than those of land plants; the supporting power of water is so great that strong stems are unnecessary. In many cases too, the large air spaces of the stem not only store and conduct air, but also increase the buoyancy or floating power of the plant.

In spite of the delicate form of water plants, many of them grow in swiftly running rivers. The flexible stems and smooth, often divided leaves stream out with the current, and are not easily torn by running water.

Flowers. Whatever the form of the plant, the flowers of practically all aquatics open above the surface of the water (Fig. 67).

Marsh plants. Marsh plants are intermediate between water
and land plants. They often occur as a fringe along the sloping
edges of ponds or rivers, growing either on wet soil or in shallow
water. Unlike most true aquatics the leaves of marsh plants are
generally above the water.

Fig. 70. THE FLOWERING RUSH. (Cambridgeshire.)

The organs of marsh plants, like those of aquatics, have large
air spaces or passages (Fig. 69 E), by means of which oxygen can
be carried to parts living in water or wet soil. No doubt the
possession of these air passages enables marsh plants to live in
wet places where ordinary land plants would die.

Among the commoner marsh plants are the Reed (Fig. 66),
Rushes, Sedges, Reedmace or Bulrush, Flowering Rush (Fig. 70),
and Iris, all of which are Monocotyledons with long, narrow
leaves. Of Dicotyledons there are the Purple Loosestrife, Water
Mint, Meadow Sweet and many others.

HINTS FOR PRACTICAL WORK.

1. Make a list of all the water and marsh plants you can find and identify, in each case noting where the plant grows, e.g. in deep or shallow water, still or running water, damp soil and so on.

2. Note also whether the plants are attached or free, and whether the leaves are submerged, floating or aerial. Make drawings of as many submerged and floating leaves as you can find.

3. Cut across the stems of a Pondweed, Water Milfoil, Reed, and other water or marsh plants, also the petiole of a Water Lily, and look at the large air spaces with a lens.

4. Many submerged water plants can be readily grown in an aquarium. A broad inverted bell-jar mounted on a block of wood makes a simple aquarium. Arrange the plants at the bottom of the jar with a little soil covered by pebbles. The water will rarely need changing if the jar is covered with a sheet of glass to keep out dust.

CHAPTER XVII

OTHER WAYS OF OBTAINING FOOD

So far we have been concerned with ordinary green plants, which make their own food from simple chemical substances absorbed from the air, soil or water by which they are surrounded. But some plants live, as animals do, on ready made food obtained either from other plants or from animals. Many of these plants have no chlorophyll, and are therefore unable to make food for themselves.

Parasites. "The ancients understood by parasites people who intruded uninvited into the houses of the rich in order to obtain a free meal." Nowadays we say that parasitic plants or animals are those which obtain food from other living plants or animals, to which they become attached.

The Dodder is a parasite with slender, thread-like stems, but neither leaves nor roots. Being sensitive to contact, it twines like a tendril round the stem of its host. The Dodder has no chlorophyll, but obtains its food by means of suckers (Fig. 71 *s*) which penetrate the stem of the plant on which it grows (often Gorse, Clover or Thyme). The Dodder is a complete parasite, that is, it gets all its food from the plant on which it grows.

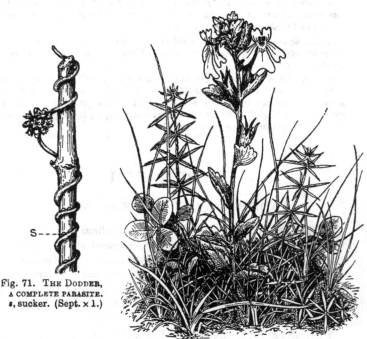

Fig. 71. THE DODDER, A COMPLETE PARASITE. *s*, sucker. (Sept. × 1.)

Fig. 72. THE EYEBRIGHT, A SEMI-PARASITE, GROWING AMONGST DWARF GRASSES, ETC. (Sept. × 1½.)

Other examples are Broomrape (feeds on Broom, Thyme, etc.) and Toothwort (on roots of trees). These three plants occur wild in Britain, but are not very common.

The Mistletoe, Eyebright (Fig. 72), Yellow Rattle and some other British Flowering Plants are **semi-parasites.** They have green leaves, and can therefore make food for themselves, but they attach themselves to roots or other parts of plants, from which they derive part of their nourishment. The Mistletoe lives on trees, but the Eyebright (Fig. 72) and Yellow Rattle prey on grasses. The bad effect that parasites have on their hosts may sometimes be seen in meadows; where the Yellow Rattle is plentiful, the grasses are stunted and less luxuriant than elsewhere.

Saprophytes are plants which, although they have no chlorophyll, do not attack living hosts, but absorb food from decaying

Fig. 73. PUFF BALLS, SAPROPHYTIC FUNGI. (Sept. × 1.)

humus in the soil (Gk. *sapros*, rotten, *phyton*, a plant). A few saprophytes are Flowering Plants, e.g. the Bird's-nest Orchid, which grows in woods; but most belong to a group of lower plants known as the Fungi, e.g. Mushrooms, Toadstools, Puff Balls (Fig. 73) and Moulds. The Fungi have neither stems, leaves nor roots, their vegetative parts consisting of delicate threads which are often underground. The conspicuous parts above ground are the fruiting bodies. Most Fungi are either parasitic or saprophytic. Saprophytes help to bring about decay of the dead bodies of plants and animals. Part of the dead matter they use for themselves, the rest being converted once more into the simple chemical substances needed by green plants.

Insectivorous plants. Most animals obtain their food by eating plants; a few plants turn the tables on the animal kingdom by capturing and devouring animals, generally small insects. These insectivorous plants have chlorophyll and can therefore make starch. Part of their food, however, is procured from the bodies of insects. Most insectivorous plants grow in boggy places which are poor in mineral salts; a few are water plants.

The Sundew (Fig. 74) is a common plant on damp moorlands, where its rosettes of reddish leaves often form conspicuous spots of colour on the spongy Bog Moss (Sphagnum). Each leaf-blade bears a number of hairs with sticky tips, which glisten like dew-drops in the sun, so that the name "Sundew" is appropriate. Small flies appear to be attracted by the "dew-drops," possibly mistaking them for honey. If a small fly alights on a leaf, it is held fast by this sticky fluid. Gradually the hairs bend over the strug-

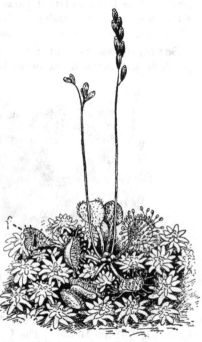

Fig. 74. ROUND-LEAVED SUNDEW (IN FRUIT), WITH BOG MOSS. *f*, leaf with fly. (Sept. Slightly reduced.)

gling fly (Fig. 74 *f*), and at the same time pour out digestive juices which slowly dissolve the softer parts of the insect. When the leaf has absorbed the nutriment it opens again, and the hard parts of the fly are blown away.

The Butterwort (Anglo Saxon *wyrt*, a plant) grows in damp places. Like the Sundew, it is common in the more rainy mountainous parts of the British Isles. It too forms a rosette of leaves,

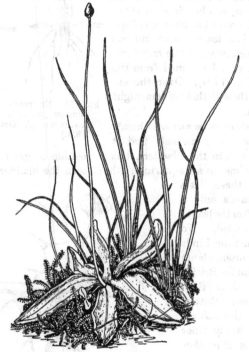

Fig. 75. BUTTERWORT (IN FRUIT), WITH MOSSES AND GRASSES.
(July, ×5/8.)

which are broader than those of the Sundew, and yellowish-green in colour (Fig. 75). The edges of the leaf are slightly curled, and its upper surface is sticky. Small insects are caught in this sticky fluid and their soft parts digested. In the autumn the leaves of both Butterwort and Sundew die away, but the terminal bud persists (Fig. 76) and forms new leaf-rosettes in the spring-time.

The Bladderwort catches its prey in a remarkable manner. It grows in still water, and has finely divided submerged leaves but no roots (Figs. 67, 77). The leaves bear curious hollow bladders with little trapdoors which can only be opened from the outside (Fig. 77 *bl*). Over the entrances to these little bladders might be written

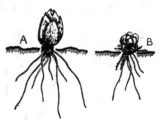

Fig. 76. BUTTERWORT (A) AND SUNDEW (B) IN THE WINTER CONDITION. (A, Oct., B, Mar. ×3/4.)

"All hope abandon, ye who enter here,"

for minute water animals find it easy to get into the bladders, but impossible to get out again. In spring-time, in an aquarium, I have seen the bladders packed full with these little water animals only a few days after the leaves had expanded. No fewer than nine kinds of insectivorous plants occur wild in the British Isles, i.e. three Sundews, three Butterworts and three Bladderworts. Many other kinds are found in other countries, including the Pitcher Plants and Venus' Fly-trap of our hothouses.

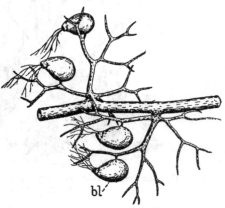

Fig. 77. PIECE OF THE SMALLER BLADDERWORT. *bl*, bladder. (Sept. ×4.)

Relations between animals and plants. We may say that green plants manufacture practically the entire food supply of the world, for they alone can make food out of simple chemical substances. All animals, as well as parasitic and saprophytic plants, depend, directly or indirectly, on green plants for food. Even carnivorous animals such as Lions and Tigers are no exception, for the animals they eat live on food made by green plants.

Another important relation which may be noticed here is the effect that animals and green plants respectively have on the composition of the atmosphere. The proportions of oxygen and carbon dioxide in the air vary remarkably little, yet during the respiration of animals and plants, and the burning of coal and other fuel, oxygen is taken from the atmosphere and carbon dioxide added to it. Were there no compensating process, the air would soon be unable to support life, for it would contain too little oxygen and too much carbon dioxide. But the supply of oxygen is constantly being renewed, and carbon dioxide removed from the air by green plants during photosynthesis. Animals then depend on green plants both for their food supply and for the air they breathe.

Other relations between animals and plants are dealt with in Chaps. XXIV, XXV and XXVII.

HINTS FOR PRACTICAL WORK.

1. In suitable places, try to find as many as possible of the plants mentioned in this chapter. Some are rare (and if found should not be disturbed), others, like the Sundew and Butterwort, are abundant, but only in certain habitats and in certain parts of the country. A few, such as Eyebright and Yellow Rattle, are very common indeed.

2. Make drawings of whole plants, or of characteristic features, e.g. the traps of the insectivorous plants.

3. The Sundew may readily be grown in a basin with wet Bog Moss. Feed one of its leaves with a very minute fragment of fresh meat, and watch the sticky hairs slowly closing over it. The Bladderwort may easily be kept in an aquarium.

CHAPTER XVIII

THE DIFFERENT FORMS OF PLANTS

I. HERBS

Herbs, shrubs and trees. There are said to be at least 136,000 different kinds of flowering plants in the world, of which something like 1800 or 1900 are natives of the British Isles. These plants show great differences of size and form. The larger plants with hard, woody, persistent stems, are either trees or shrubs. **Trees** are generally taller than shrubs, and have a single main stem or trunk, with branches well above the ground (Fig. 5). **Shrubs** on the other hand are usually

Fig. 78. HAZEL IN WINTER CONDITION; FROM A WOOD. (March.)

branched down to their base (Fig. 78). **Herbs** are smaller plants, with comparatively soft (i.e. not woody) stems which generally die down to the ground in autumn.

Fig. 79. HAREBELL WITH UPRIGHT STEM, WHICH DIES AWAY AFTER FLOWERING. The creeping shoot with round leaves (A) will live through the winter, and then grow into an erect flowering stem. (Aug. × 3/8.)

Length of life. Trees and shrubs are **perennials,** that is, they live for a number of years; some trees may even last for centuries. Many herbs also are perennial (e.g. Daisy), but others are either **annuals,** living for one year or less (e.g. Groundsel and Chickweed), or **biennials,** which last for two seasons (e.g. Foxglove and Carrot).

Herbaceous plants. Herbs show a greater variety of form or **habit** than woody plants. Most have upright stems, like the

Fig. 80. HAREBELL; DIFFERENT SHAPES OF LEAVES FROM THE SAME PLANT. (Aug. × 1/2.)

Groundsel (an annual, Fig. 1) and Harebell (perennial, Fig. 79). In these two plants and many others the lower leaves differ in shape from the upper ones (Figs. 3, 80). In biennial herbs such

Fig. 81. ROSETTE OF DAISY. A, seen from above; B, from one side. (Sept. × 1/2.)

as the Foxglove and Mullein, the stem remains short during the first summer and winter, the leaves forming a rosette just above the ground level. In the second year a tall upright flowering stem is formed. Some perennial plants keep the rosette form all their lives, e.g. Daisy (Fig. 81), Dandelion and Plantains. The

leaves of such **rosette plants** are sometimes closely pressed to the ground, so that they receive plenty of light, and at the same time prevent other small plants from overcrowding them.

EXP. 30. Dig up a Daisy rosette and keep it for a night in a closed tin. The plant curls up as in Fig. 82, because the upper surfaces of the leaves tend to grow faster than the lower. In nature this tendency helps to keep the leaves flat against the ground.

Fig. 82. DAISY, AFTER
A NIGHT IN A CLOSED
TIN. EXP. 30. (Sept.
× 1/2.)

Climbing plants (Chap. XIII) are only upright if they find a suitable support, otherwise they trail along the ground. True **creeping plants** often resemble climbers in having slender stems and long internodes, but they creep on

the surface of the ground instead of climbing. Their thin creeping stems are called **runners,** e.g. Strawberry. Runners are usually formed as axillary branches on the main stem. They spread rapidly over the ground, taking possession of any vacant space, or even crowding out less aggressive plants. Some of the buds on the runners develop into new plants, which become independent of the parent if the runner itself dies away. In this way the number of individual plants may be increased by **vegetative propagation,** that is, reproduction not by seeds, but by the growth

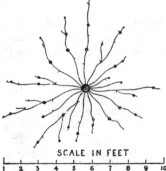

SCALE IN FEET

Fig. 83. DIAGRAM OF THREE YEAR
OLD STRAWBERRY PLANT, WITH
RUNNERS AND YOUNG PLANTLETS
(black dots).

of some vegetative part. The Strawberry often forms a great many new plants in this way (Fig. 83).

Other plants have thick, fleshy, creeping stems called **rhizomes,**
e.g. Iris and Solomon's Seal (Fig. 84). Rhizomes are usually
underground; their leaves, which are unable to make starch in
the dark, are reduced to scales. The presence of scale leaves with
buds in their axils shows that rhizomes are underground stems
and not roots. The growth of most rhizomes is sympodial (cf.

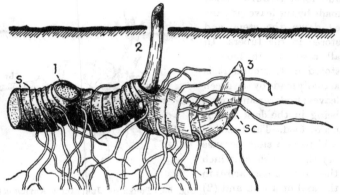

Fig. 84. Sympodial Rhizome of Solomon's Seal. 1, scar of last year's
aerial shoot; 2, base of this year's shoot; 3, terminal bud of next
year's shoot. *sc*, scale leaves; *s*, scars of old scales; *r*, adventitious
roots. (Sept. × 5/8.)

p. 76); each year the tip of the rhizome turns up and forms
leaves and flowers above the surface, the growth of the rhizome
itself being continued by an axillary bud (Fig. 84).

Food storage. Most plants store part of the food they have
made for the future use of themselves or their offspring. As a
rule annuals store food only in their seeds, while biennials and
perennials store it in their vegetative parts as well. Trees and
shrubs store much of their reserve food above ground, in their
persistent stems, but in herbs it is usually found in swollen, underground **storage organs.** These organs may be stems, leaves or
roots.

Stems which store food. Examples of underground stems which store food are the rhizomes of Iris and Solomon's Seal (Fig. 84), the **tubers** of Potato and Jerusalem Artichoke (Fig. 85), and the **corms** of Crocus and Montbretia (Fig. 86).

In the Jerusalem Artichoke some of the buds at the base of the stem grow into underground branches, the tips of which swell and form **tubers.** Food made by the leaves is passed down the stem and stored in the tubers. In all cases where food is stored underground it is actually made by the green leaves; underground parts being in the dark cannot make food. The tuber is said to be a stem because (1) the branch of which the tuber is part arises in the axil of a leaf, and (2) the tuber itself bears scale leaves and ends in a terminal bud (Fig. 85).

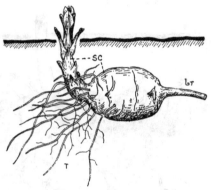

Fig. 85. TUBER OF JERUSALEM ARTICHOKE. *br*, tuber-bearing branch; *sc*, scale leaves; *r*, adventitious roots. (May, × 1/3.)

In the spring-time the terminal bud grows above the ground into a new upright stem. Potato tubers are formed in almost exactly the same way as those of the Jerusalem Artichoke. The "eyes" of the potato represent scale leaves with their axillary buds. In one season a Potato or Jerusalem Artichoke plant may give rise to many new tubers, each of which can form a new plant. This is another example of vegetative propagation.

In Crocus and Montbretia (Fig. 86) the lower part of the stem swells into a solid, rounded structure called a **corm.** A corm is a kind of upright stem-tuber, packed with reserve food material. The corm of Montbretia consists of several internodes covered by the remains of leaves (Fig. 86 A, C). The pointed bud at the tip of the corm becomes a flowering stem, while the buds in the axils of the scales may either form new leafy shoots or short creeping stems (Fig. 86 *cs*). New corms are formed either (1) from

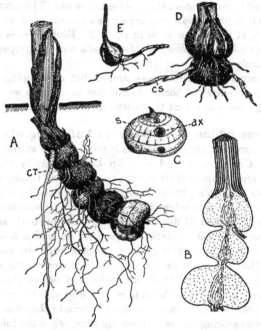

Fig. 86. CORMS OF MONTBRETIA. A, chain of corms; *cr*, contractile root. B, corms cut in half. C, corm with leaves removed; *s*, leaf-scar; *ax*, axillary bud. D, corms with creeping stems (*cs*). E, young corm formed from creeping stem. (Aug. A, D ×1/2; B, C, E ×1.)

the tips of these creeping stems (Fig. 86 E), in much the same way as the tubers of the Jerusalem Artichoke, or (2) by the swelling of the base of the leafy shoot on the top of the old corm. In the latter case a chain of corms is gradually formed (Fig. 86 A), for the old corms may live for years. Adventitious roots are formed at the base of the new corm. Some of these are larger than the others and contract (i.e. become shorter and thicker) when they have anchored themselves in the soil. The **contractile roots** (Fig. 86 *cr*) pull the corms downwards, and so keep the new ones at the proper level in the soil. Each corm is capable of an independent existence, so Montbretia too can propagate itself vegetatively.

The corms of the Crocus are similar, but no creeping stems are formed; also the old corms do not last so long, so we never find the long chains of corms that occur in Montbretia.

Leaves which store food. In **bulbs** (Fig. 87), which resemble corms in appearance, the reserve food is stored mainly in leaves. Cut a corm and a Tulip bulb vertically through the middle, and compare their structure. The corm is almost entirely made up of a solid, food-storing stem, while the bulb consists of a short, flat stem bearing three or four fleshy, food-storing leaves (cf. Figs. 86 B and 87 B). In the centre of the bulb are a small erect stem, several miniature foliage leaves and a flower which would expand next spring. Cut another bulb transversely and notice that the scale leaves completely surround the erect stem (Fig. 87 C). Remove the fleshy scales and notice that each one has a bud in its axil. For this reason, and also because they are borne laterally on a stem, we may conclude that the scales are really food-storing leaves. In spring-time, when the foliage leaves are making food, some of it travels down to the axillary buds, which gradually develop into new bulbs. The whole Tulip bulb is really a bud, because (1) like other buds (cf. p. 122 and Fig. 92) it consists of a short stem, with crowded nodes and overlapping leaves, and (2) it arises in the axil of a leaf.

In the Tulip the bulb scales are complete scale leaves, but in the Onion, Bluebell and Daffodil (Fig. 87 D) they are merely the swollen bases of foliage leaves, which persist after the leaves themselves have withered.

Each parent bulb may produce several daughter bulbs which become separated from one another on the death of the parent. Bulbous plants propagate themselves vegetatively by this means.

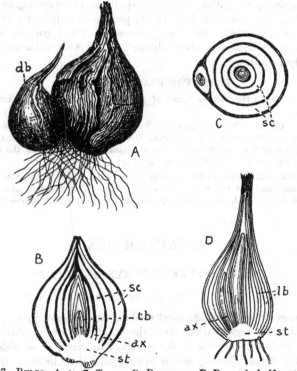

Fig. 87. Bulbs. A, B, C, Tulip; D, Daffodil. B, D, cut in half vertically; C, transversely. *st*, stem; *sc*, fleshy scale leaves; *tb*, terminal bud with leaves and flower; *lb*, fleshy bases of foliage leaves; *ax*, axillary bud. The two bulbs in A are sister bulbs of the same age. (Sept. × 3/4.)

Roots which store food. Food may also be stored in tap-roots, e.g. Parsnip, Turnip and Carrot, or in tuberous adventitious roots, e.g. Dahlia and Lesser Celandine (Fig. 105).

Use of stored food. The food which herbaceous plants store in various underground parts is used for the following purposes: (1) It enables the plant, or those parts of it which are used for vegetative propagation, to grow rapidly in early spring, or even in winter (cf. Chap. XXI). (2) It provides material for respiration, which releases the energy necessary if the growing shoots are to force their way above the soil, or the roots to burrow more deeply into it.

HINTS FOR PRACTICAL WORK.

1. Examine the various forms of plants mentioned, while reading this chapter.

2. Plant some Crocus corms and Tulip or other bulbs in pots or in the garden, and study their life-histories. Draw a Crocus corm, which has been cut vertically through the middle, as it appears in the autumn, the spring, and the summer (before the leaves die away).

3. Cut pieces from various storage organs—e.g. corms, bulbs, tubers and rhizomes, and test the cut surfaces with iodine. You will find that starch (which turns bluish-black with iodine) is a common form of reserve food.

CHAPTER XIX

THE DIFFERENT FORMS OF PLANTS

II. TREES AND SHRUBS

Trees are amongst the most striking and interesting of plants, whether we regard them from the point of view of their great size, or the usefulness of their timber to man; or consider the beauty of their fresh greens in spring-time, their brilliant tints in autumn, and the delicate tracery of their bare twigs in winter.

Tree trunks. The main stem of the seedling of a tree is no thicker than that of a herb, but year by year it grows till it becomes a massive trunk, often several feet in diameter. Look at the sawn end of a trunk or branch of a tree which has been cut down. It is hard and tough, and consists for the most part of a series of concentric rings of woody tissue. One new ring is formed each year, just outside those of previous years. By counting the number of these **annual rings,** as they are called, we can find out the age of the trunk or branch. The wood is made up partly of water-conducting vessels (p. 48) and partly of tough fibres which strengthen the stem, and enable it to support the weight of the numerous branches and leaves. Outside the wood is a narrow band of tissue, the inner part of which serves to conduct food from the leaves to other organs of the plant, while the outer part forms the **bark** of the tree. The bark consists mainly of **cork,** a waterproof substance which prevents the escape of water even more effectively than does the skin of a herbaceous plant. On the bark of a young twig small round or oval spots (the lenticels, Figs. 88, 91) are seen. The **lenticels** are porous places through which air can pass, the stem like other parts needing oxygen for respiration. In a bottle cork, which is made from the thick bark of the Cork Oak, the lenticels appear as streaks or spots filled with a brown powdery substance. A good bottle cork is always cut so that the lenticels run across the cork ; if they ran lengthwise, liquid might escape through them. As the stem increases in thickness, the bark becomes too small, and tends to split, new layers of cork being formed underneath the old ones. In most trees the bark cracks but does not fall off, so that the trunk has a rugged, ridged appearance. This type of bark is known as **fissured bark,** e.g. Oak and Elm. The Sycamore and Plane have **scaly bark,** the older bark falling off in scales or flakes. The bark of the Birch peels off in broad ring-like papery pieces (**ring bark**). Thick-barked trees such as the Oak can stand exposure better than the Beech and other trees with thinner bark.

Twigs of trees. The twigs or smaller branches are well worth studying, especially in the winter. Fig. 88 shows several winter twigs of the Sycamore. In A, a strong vigorous twig, notice the following: (1) the terminal bud; (2) several pairs of

entered the leaf from the stem; (3) numerous lenticels in the bark of the internodes; (4) two lateral branches each with a leaf-scar below it, and (5) just above the branches, a girdle of narrow scars (*g*).

In the spring-time the terminal bud will unfold (Fig. 8) and grow into a new length of leaf-bearing stem. As the bud opens, the scales fall off, leaving a girdle of scars round the stem, like those in Fig. 88 *g*; the girdle therefore marks the position of an old terminal bud. This process is repeated every year, a new terminal bud being formed, which rests during the winter and un-folds in the following spring. The portion of stem between the two girdles therefore represents one year's growth, so by counting the number of girdles we can find the age of a twig. The actual length of a year's growth varies very much; A (Fig. 88) repre-sents less than two years' growth, while B, a twig which has been shaded, and therefore starved, by more vigorous branches, took eight years to grow six inches. I once found a Sycamore twig which had grown 31 inches, and an Ash twig 45 inches in one year.

Branching of twigs. Now examine twig C (Fig. 88), which has a forked appearance, very different from that of A and B. In each fork there is a saddle-shaped scar (*inf. s*), caused by the dying away of the fruiting stem (cf. D, *inf.*), for in this case the terminal bud gives rise to the inflorescence, and then ceases growth. Next year the two nearest axillary buds grow into branches, and so the stem forks. We thus find two methods of branching in Sycamore twigs, (1) the **monopodial** method of the vegetative twigs (Fig. 88 A, B), in which the tip of the stem grows on in a straight line from year to year, the axillary branches being shorter than the main one; and (2) the **sympodial** method of the flowering twigs (Fig. 88 C, D), where the tip of the stem dies away, growth being continued by means of axillary branches, which overtop the main one.

In the Lime (Fig. 89) the vegetative twigs are sympodial. The apparently terminal bud (A, *ax*) is really axillary, as will be seen by reference to B and C. During the summer the real terminal bud (C, *tb*) dies away leaving a rounded scar (B, *b.sc*). The foliage leaf falls in the autumn, but its scar (*fs*) remains, and the bud in its axil continues the growth the following year. The branching is sympodial, and a twig of several years' growth is made up of a series of joints, each of which is really a lateral branch formed from an axillary bud.

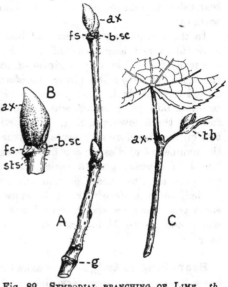

Monopodial branching occurs in the Sycamore, Horse Chestnut etc., and in Pine and Fir trees. Most of our trees, however, branch sympodially, as also do the Virginian Creeper

Fig. 89. SYMPODIAL BRANCHING OF LIME. *tb*, true terminal bud; *ax*, apparently terminal, but really axillary bud; *b.sc*, scar of terminal bud; *fs*, scar of foliage leaf; *sts*, scar of stipule; *g*, girdle scar. (A, B, Feb.; C, May. A, C ×3/4; B ×1½.)

(Fig. 50) and many herbaceous plants with rhizomes (Fig. 84).

History of twigs. By carefully observing winter twigs we may learn a good deal about their history, both past and present. For instance, the twig shown in Fig. 90 was taken from a Beech hedge. For four years the lateral branch grew slowly, but much more quickly in the fifth year (note positions of girdles) owing to the cutting off of the main branch (Fig. 90 *cb*). This enabled the lateral branch to get more food and water, and so to grow rapidly.

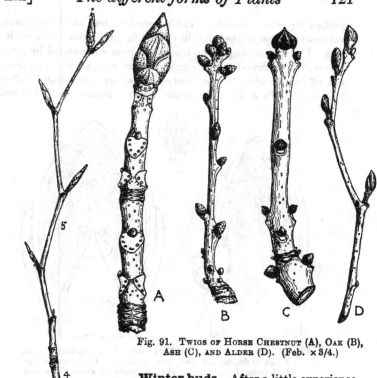

Fig. 91. Twigs of Horse Chestnut (A), Oak (B), Ash (C), and Alder (D). (Feb. × 3/4.)

Fig. 90. Twig of Beech. 1–5, portions formed in successive years. *cb*, cut end of main branch. (Feb. × 5/8.)

Winter buds. After a little experience it is easy to recognize most of our common trees by the size, shape and colour of their winter buds. The large dark brown, sticky buds of the Horse Chestnut, for example, are quite distinct from the stumpy black buds of the Ash, or the long, narrow, pointed, pale brown buds of the Beech (cf. Figs. 88 to 91).

The buds are covered with overlapping scales, which are hard and tough and tightly

packed together. Carefully cut a terminal bud of the Sycamore
vertically through the middle (Fig. 92), and examine it with a
lens. The bud is simply the tip of the stem surrounded by young
leaves, the internodes between which are very short. The leaves
are of two kinds; in the centre are a few very small foliage
leaves (Fig. 92 A, c), and outside these a number of hairy bud-

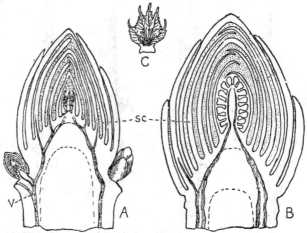

Fig. 92. TERMINAL BUDS OF SYCAMORE. A, a vegetative bud;
B, an inflorescence bud, cut vertically. C, young foliage
leaf from A. *sc*, bud-scales; *v*, veins of stem. The dots
represent hairs. (Feb. A, B × 5; C × 10.)

scales. Some of the larger terminal buds contain flowers as well
as leaves (Fig. 92 B). We see then that the leaves and flowers
which grow so rapidly in spring-time are actually formed the
previous year.

Nature of bud-scales. Bud-scales are very different from
foliage leaves, but none the less they represent either leaves or
parts of leaves. The bud-scales of Lilac and Privet are *leaf-blades*,
as may be seen in an opening bud, where stages intermediate

between scales and foliage leaves are found (Fig. 93). In the Sycamore, Ash and Flowering Currant the scales are *well developed
leaf-bases.* These scales often have a small leaf-blade, and some-

Fig. 93. NEWLY OPENED BUD OF LILAC; THE BUD-SCALES
ARE LEAF-BLADES. (Apr. × 3/4.)

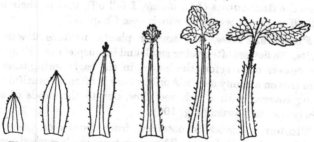

Fig. 94. FLOWERING CURRANT; THE BUD-SCALES ARE LEAF-BASES.
(Apr. × 1½.)

times a short petiole as well, which proves that the scale itself is
really a leaf-base. Such intermediate forms are regularly found in
the Flowering Currant (Fig. 94). Finally, in many trees, such as

the Beech, Alder and Lime, the bud-scales are *stipules*, arranged in pairs on either side of the foliage leaves (Fig. 95). The nature of bud-scales can best be studied in spring-time, when the buds are unfolding. When the scales fall, only their scars remain to mark the place where they had been.

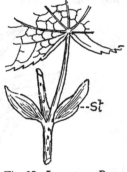

Fig. 95. LIME; THE BUD-SCALES ARE STIPULES (*st*) WHICH FALL WHEN THE BUD OPENS. Cf. Fig. 89, *sts*. (Apr. × 7/8.)

Foliage leaves. Most of our common trees and shrubs belong to the Dicotyledons, their leaves having the net-like arrangement of veins characteristic of that group of plants. The leaves themselves are of various shapes, like those of herbs; in a few cases (e.g. Horse Chestnut, Ash, Elder) they are compound, but most are simple, with entire, toothed or lobed edges (cf. Figs. 7, 40 A, E, 43, 93). A number of familiar trees are not Dicotyledons but Conifers, a group of plants with needle-shaped leaves (Fig. 38 B). Conifers, e.g. Pines, Firs and Cedars, are usually **evergreen,** but most of our Dicotyledonous trees are **deciduous** (Lat. *decido*, I fall off), that is, their leaves fall off on the approach of winter (see Chap. XX).

Flowers. Trees, like smaller plants, produce flowers and fruits, the flowers often being small and inconspicuous. Frequently the flowers open before the leaves, in the early spring-time. An Elm tree on a sunny day in April or May is often a beautiful sight, being covered with a rich red glow, due to thousands of small, newly opened flowers (Fig. 100).

Timber. Timber is obtained from trunks of trees, both Conifers and Dicotyledons. The softer, more easily worked timbers, such as deal and pitch pine, are usually coniferous, the harder woods, e.g. oak, mahogany, walnut, being obtained from Dicotyledons.

The life-processes of trees. The numerous roots of a tree which, if placed end to end, might measure miles in length, absorb water and mineral salts, and its myriad leaves trap sunshine, make food, and transpire just like those of the humble herbs growing at its foot. Owing to the great number of its leaves, a tree transpires enormous quantities of water, and therefore needs to absorb a great deal from the soil. For this reason woods and forests are only found in places where the subsoil contains a considerable amount of water, and in countries where the rainfall is fairly high. The climate of most parts of the British Isles is suitable for the growth of woodland, and in pre-Roman times forests were far more extensive than at the present time.

Undergrowth. Just as we find that only certain plants can live in water (see Chap. XVI), so only certain kinds can live under trees. Trees absorb so much light that the undergrowth in many woods is composed for the most part of such shade-enduring plants as Bluebell, Wood Sanicle, Wood Sorrel, Wood Anemone, Ferns and Mosses. Some of these woodland herbs will be considered in another chapter.

The forms of plants. We have seen that Flowering Plants may differ greatly from one another in size, form and appearance. But they are all built up on the same general plan. All have roots, stems and leaves. Stems are made up of nodes and internodes, leaves are borne laterally on stems (the youngest always being near the tip), and branch shoots arise in the axils of leaves. You may perhaps have found the terms shoot, stem and branch a little confusing. A **shoot** is a leafy stem, that is, it consists of leaves as well as stem. A **stem** is the part of the shoot on which the leaves are borne. A **branch shoot** is any part of the shoot-system which arises from a bud in the axil of a leaf.

HINTS FOR PRACTICAL WORK.

1. It adds interest to a walk to be able to recognize the common trees. A good plan is to start with one or two trees such as Sycamore, Oak, Ash or Beech, and get to know them thoroughly at all seasons of the year. Examine, describe and draw carefully their leaves (noticing shape, veins,

etc.) and winter twigs (noticing shape of buds, leaf-scars, lenticels, method of branching, etc.), and learn to recognize these particular trees by their bark. When you know these trees really well, you can study one or two new ones each year in the same way.

2. Examine winter twigs of Sycamore and Lime. Collect twigs of Lilac, Horse Chestnut, Beech and Elm, and find out for yourself their methods of branching.

3. In the spring-time examine the opening buds of Lilac, Sycamore, Flowering Currant, Lime, Alder and Beech, and find out the nature of their bud-scales.

4. Select some particular woodland, and make a list of the plants forming the undergrowth. Compare this list with similar lists of plants found in meadows, hedgerows, water, etc. In this way you will gradually learn which plants grow in particular habitats.

CHAPTER XX

HOW PLANTS PASS THE WINTER

I. Trees and Shrubs

Summer and winter. Everybody knows that most plants grow and flower in spring or summer. Winter is often looked upon as a season when plant life is at a standstill. This is not altogether true, though the short days of winter are certainly far less favourable for plants than the longer days of summer. This is because during the winter the temperature is lower and there is less light for photosynthesis. Plants are greatly affected by temperature, and it has been found (1) that warmth is necessary, not only for growth (Exp. 2, p. 30), but also for such processes as the absorption and transpiration of water; and (2) that broadly speaking, the higher the temperature, the more rapidly are these processes carried on. These facts should be borne in mind in studying the different ways in which plants endure the hardships of winter.

Trees and Shrubs in Winter.

Leaf-fall. A remarkable thing about woody plants is that so many of them, especially those with broad thin leaves, shed their leaves in autumn. Is this curious habit of any advantage to the plant? In order to attempt to answer this question it is necessary to remember, (1) what has been said above about temperature, (2) that most trees send their roots deep into the soil, and (3) that the temperature of soil changes much more slowly than that of air. Take the case of a tree exposed in winter or

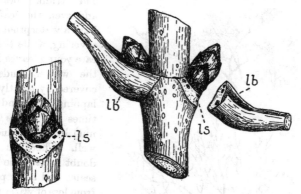

Fig. 96. LEAF-FALL IN SYCAMORE. *lb*, leaf-base; *ls*, leaf-scar.
(Nov. × 2.)

early spring to a sudden spell of warm weather after a long frost. The ground has gradually become colder and colder till the roots are no longer able to absorb water from the frozen soil. If the tree had retained its leaves, they would begin to transpire as soon as the air became warmer. The roots, however, being deep in the cold soil, could not absorb enough water to replace that lost by transpiration, and the tree might die. The autumnal shedding of leaves, then, probably lessens the danger of over transpiration in winter or early spring, when the air is warm but the soil cold.

Before the leaf falls, a layer of cork is formed across the inside

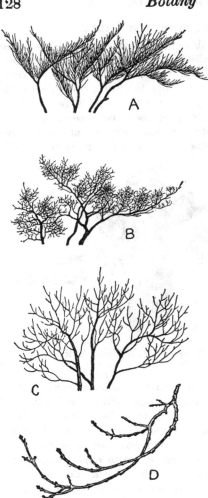

Fig. 97. SMALLER BRANCHES OF BEECH (A), ALDER (B), ASH (C). D, A LOWER BRANCH OF AN ASH TREE. (March.)

of the leaf-base, so that when the leaf is separated from the twig the scar is already healed (Fig. 96). Every part of the tree which is above ground is now protected against too great loss of water during the winter. The trunk, branches, and even the leaf-scars have a waterproof corky covering, while the delicate young leaves inside the winter buds are covered by tightly overlapping scales, and sometimes (e.g. Horse Chestnut) by woolly hairs as well. There is little doubt that the bud-scales and hairs protect from loss of water rather than from cold. The buds may be completely frozen without being injured, but if the scales are removed, even in warm weather, the delicate leaves inside soon wither and die.

Recognition of trees. In the bare winter condition many kinds of trees may readily be recognized, even at a distance. Notice the differences between the smaller twigs of the Sycamore, Beech, Alder and Ash (Figs. 5, 97). In several trees the tips of the lower branches tend to turn upwards; this is especially characteristic of the Ash (Fig. 97 D).

Pruning. Fruit growers prune their trees in order to remove unnecessary branches, and to stimulate the growth of flowering shoots. In nature, trees not only remove their own leaves, but in some cases superfluous twigs as well, by a kind of **natural self pruning,** e.g. Black Poplar and Oak (Fig. 98).

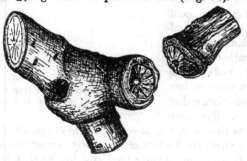

Fig. 98. SELF PRUNING OF OAK. (Dec. × 1½.)

Evergreens. In some trees and shrubs the leaves last for several seasons instead of only one. Such plants are known as **evergreens,** because they have green leaves all the year round. The leaves of evergreen trees and shrubs usually have a very thick waterproof skin, which gives the leaf a tough, leathery texture (e.g. Holly, Fig. 36, Rhododendron). Often too the leaves are small, as in Conifers, most of which are evergreen. Such leaves do not transpire so rapidly as the thinner, more delicate leaves of deciduous trees. Because of the structure of their leaves, evergreens are not exposed to the risk of excessive loss of water by keeping their leaves through the winter.

Opening buds. In spring-time, when the soil is warm enough for the roots to absorb water, the buds begin to swell and elongate.

Soon the buds open and the young leaves appear. Trees often burst into leaf with surprising rapidity. This is possible because the little leaves and flowers were formed in the bud the previous summer. They rest during the cold of winter and expand when stimulated by the warmth of spring. Opening buds often take up a more or less vertical position, either erect, e.g. Sycamore

Fig. 99. OPENING BUDS OF BEECH, IN NATURAL POSITIONS. *st*, stipular bud-scales. (May, × 5/8.)

(Fig. 8), or drooping, e.g. Beech and Elm (Figs. 99, 100). In the Horse Chestnut the leaflets of the newly opened compound leaves droop before taking up their final position, so that the leaf looks like a half-opened umbrella (Fig. 61). Some trees come into leaf much earlier than others. The Hawthorn is one of the earliest, then such trees as the Horse Chestnut, while the Oak and Ash are always late.

"If the Oak comes out before the Ash, we shall only have a splash,
But if the Ash comes out before the Oak, then indeed we'll have a soak."

HINTS FOR PRACTICAL WORK.

1. Learn to recognize at a distance the particular trees selected (see Pract. Hints, Chap. XIX), by means of their finer twigs (cf. Fig. 97).

2. Examine trees in autumn, when shedding their leaves. Gently pull off some leaves, and notice the smooth leaf-scars, already covered with cork.

3. Examine during the winter Beech and Oak trees of various ages. You may notice that in young trees the leaves die in autumn, but remain on the tree for a long time. When do these dead leaves finally fall from the trees? What happens to the leaves of older trees?

4. Make a list of different kinds of evergreen trees and shrubs you come across. How many of them have thick, leathery leaves? Are their leaf-edges smooth (entire) or toothed or lobed?

5. In spring-time, notice which trees unfold their leaves early and which are late.

6. Examine opening buds of Sycamore, Elm, Lime, Ash, Beech and Horse Chestnut. Which have erect and which have drooping buds or young leaves?

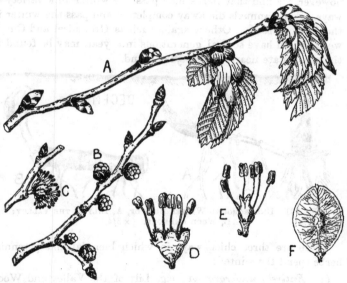

Fig. 100. ELM. A, opening buds. B, unopened, and C, opened flowers D, single flower. E, ditto with perianth removed. F, young fruit (A—E, Apr., F, May. A × 1¼, B, C, F × 5/8, D, E × 2¼.)

CHAPTER XXI

HOW PLANTS PASS THE WINTER

II. Herbs

Unlike woody plants, most herbs die down in autumn, nearly or quite to the level of the ground. If we study them carefully, however, we find that herbs may pass the winter in a variety of ways. Some annuals die away completely and pass the winter in the form of seeds. Others again, such as Groundsel and Chickweed, which have several generations in a year, may be found in the green state nearly all the year round.

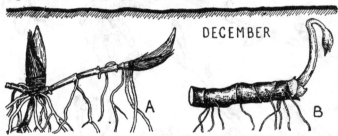

Fig. 101. Underground Winter Buds of, A, Lily of the Valley; B, Wood Anemone. (× 3/4.)

There are three chief ways in which biennial and perennial herbs spend the winter:

(1) *Entirely underground;* e.g. Lily of the Valley and Wood Anemone (Fig. 101). When the time comes, next year's shoots, like those of seedlings, may force their way to the surface either by means of a stiff, spear-like point, or by a bent loop, which protects the delicate, growing tip (cf. Figs. 21, 23, 101).

(2) *With buds just on the surface of the soil, but with no green*

leaves; e.g. Meadow Sweet and Gout Weed, Great Plantain, Sundew and Butterwort (Figs. 102, 103 A, 76).

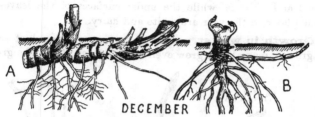

DECEMBER

Fig. 102. SURFACE WINTER BUDS OF, A, MEADOW SWEET;
B, GOUT WEED. (× 1/2.)

(3) *With green leaves just above the surface of the ground;* e.g. most grasses and many rosette plants, such as Daisy, Ribwort Plantain, Lesser Spearwort (Figs. 81, 103 B, 104) and Foxglove.

Forms of leaves. In some cases herbaceous plants have different kinds of leaves at different times of the year. E.g. the winter and summer leaves of the Lesser Spearwort are quite

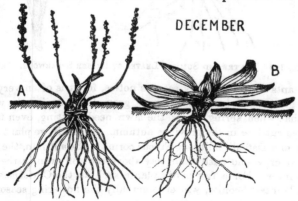

DECEMBER

Fig. 103. A, GREAT PLANTAIN, WITH BUD ON THE SURFACE;
B, RIBWORT PLANTAIN, WITH GREEN LEAVES. (× 1/2.)

different in shape (Fig. 104). Again, the first spring leaves of the Meadow Sweet (unfolded about February or March) are green and hairless, while the under surfaces of the leaves unfolded later in the year are white and hairy.

Growth in winter. Although summer is the chief season of growth, many herbs grow even in the winter-time. The growth

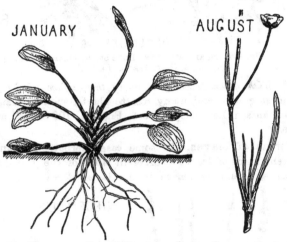

JANUARY   AUGUST

Fig. 104. WINTER AND SUMMER LEAVES OF LESSER SPEARWORT. (× 1/2.)

of plants in winter easily escapes notice, for it is often very slow. The grasses on a lawn, for example, grow slowly during the winter, so that when spring comes the lawn needs cutting, even though it was cut late in the previous autumn. Again, if we plant Snowdrop and Daffodil bulbs, or Crocus corms in the autumn, the young shoots grow and push their way above the soil even in the depth of winter. For a few hardy plants such as the Christmas Rose and Winter Aconite, winter is actually the flowering season.

Woodland herbs. Many of the spring flowering herbs of our woods and hedgerows grow a good deal during the winter,

and a wood is often gay with flowers before the trees awaken
from their winter rest. In the summer-time the smaller plants
are shaded by the foliage of the trees under which they live, so
that early spring, before the trees burst into leaf, is perhaps their
most favourable season. It is interesting to notice how these
plants prepare for the spring flowering season. They lose their

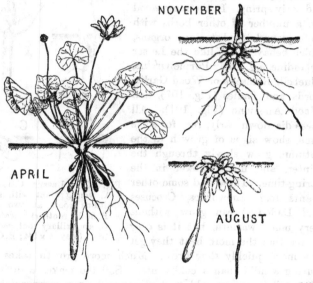

Fig. 105. LESSER CELANDINE AT VARIOUS SEASONS OF THE YEAR. (× 1/2.)

leaves and enter on the resting period much sooner than trees or
summer flowering herbs, for they have to make a very early start.
For instance, in a wood near my house, the Lesser Celandine (see
Fig. 105, April) often dies away as early as May or June. For
the next two or three months the only living parts of the plant
are a cluster of food-storing, adventitious roots (Fig. 105, August),
or similar roots which are often formed from buds in the leaf-
axils (Fig. 106, B). About September a young green shoot becomes

visible, and this grows slowly (Fig. 105, November) until by about Christmas time several leaves have opened, and even flower buds may in some cases be seen. By March the plants may be in full flower. This plant grows chiefly during the milder periods of winter and early spring. In the same wood are a number of other herbs with underground food-storing organs, which behave much like the Lesser Celandine, e.g. Early Purple Orchis, Bluebell, Ramsons or Wood Garlic, Lords and Ladies (Fig. 107), and Wood Anemone (Fig. 101). All these die down early, rest for some time, show signs of growth in the autumn, grow slowly through the winter, and finally flower in the spring-time. These and some other plants (e.g. Snowdrops, Crocuses and Daffodils) can grow without very much warmth, but it is easy to see that the more heat they get

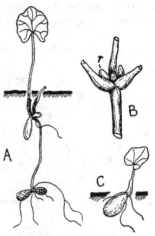

Fig. 106. LESSER CELANDINE. A, young plant with adventitious roots at two levels. B, fleshy roots (*r*) in leaf-axils. C, an axillary root germinating. (May, A × 3/4; B, C × 1½.)

the more quickly they grow. Much more growth takes place during a mild than a cold winter. Soil too makes a difference. Wet soil is always cold, and plants will grow more quickly on a dry, light soil than on a wet, heavy one. The Lesser Celandine, for example, will often be in full flower on a dry, sunny bank, on the south side of a hedgerow, when its less fortunate fellows on the damp, shady floor of a neighbouring wood have no open flowers at all.

The roots of small plants are usually nearer the surface of the ground than those of trees, and are more easily warmed by the sun. This is no doubt one reason why woodland herbs so often grow while the trees are still in the resting condition. Then too we must remember that many woodland herbs accumulate large stores of food before the resting period begins. This stored food makes it possible for them to grow during the winter when the light is dull, and fresh food cannot be made so easily as in summer. The food is stored underground in bulbs (Bluebell, Ramsons), corms (Lords and Ladies, Fig. 107), rhizomes (Anemone, Fig. 101), or roots (Orchis, Lesser Celandine, Fig. 105).

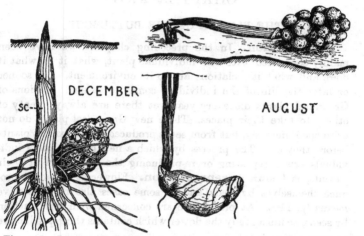

Fig. 107. LORDS AND LADIES. August, resting corm; stalk of the red berries has decayed. By December, bud of corm has grown considerably. *sc*, scale leaf. (× 1/2.)

Frost. Frost does not as a rule injure our hardy native plants, but it may do so if the frost occurs in late spring, after the plants have begun to grow, or if it is accompanied by strong, drying winds. On the other hand, some of the less hardy garden plants introduced from warmer countries, such as Potato, Vegetable Marrow and Garden Nasturtium, are readily killed by frost.

HINTS FOR PRACTICAL WORK.

1. During the winter-time try to find as many as possible of the common wild herbaceous plants you know, and compare their winter states with those mentioned in this chapter.

2. A good plan is to study carefully some one plant right through the year. E.g. Great Plantain, White Dead Nettle, Lesser Celandine or Marsh Marigold. Make notes and drawings of the seasonal differences observed.

3. Examine and draw the underground storage organs of some common woodland plants, e.g. Bluebell, Orchid and Wood Anemone.

CHAPTER XXII

THE FLOWER OF THE BUTTERCUP

Reproduction. In the preceding chapters we have been mainly concerned with the individual plant, what it is, what it does, and what its relations are to its environment. But sooner or later the life of the individual comes to an end. Millions of Groundsel plants die every year, yet there are always plenty of others to take their places. These new Groundsel plants do not arise spontaneously, but from seeds produced by the parent plants before they die. The process by which a new generation of individuals arises, repeating or re-producing the characters of their parents, is known as **reproduction.** Flowering Plants reproduce themselves by seeds, and in some cases also by vegetative means (p. 110). At present we are concerned with reproduction by seeds, so must study the flower which contains the reproductive organs.

The flower of the Creeping Buttercup. Examine a Buttercup flower, and notice that it is made up of the same kinds of organs as the flowers described in Chap. II. The top of the flower-stalk is the **receptacle** of the flower, and on the receptacle are closely set four kinds of **floral leaves,** i.e. :

(1) Five green, leaf-like **sepals** (Fig. 108 A), arranged in a circle or **whorl** on the outside of the flower. The five sepals together form the **calyx** (Gk. *kalyx,* a cup).

(2) Just inside the calyx is a whorl of five bright yellow **petals** (Fig. 108 B), which together form the **corolla** (Lat. a little crown).

At the base of each petal is a small pocket-like honey gland or **nectary.**

(3) Next we find a number of yellow **stamens,** each with a thin stalk, the **filament,** at the tip of which is a swollen **anther** (Fig. 108 c).

(4) In the centre of the flower are a number of minute green structures, the **carpels,** each of which consists of an **ovary, style** and **stigma** (Fig. 108 D). In each ovary is a single **ovule** (Fig. 110).

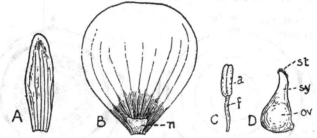

Fig. 108. CREEPING BUTTERCUP. A, sepal. B, petal; *n,* nectary. C, stamen; *a,* anther; *f,* filament. D, carpel; *ov,* ovary; *sy,* style; *st,* stigma. (July, A—C × 4½; D × 9.)

Recording observations. After carefully examining the structure of the Buttercup flower, you should record your observations by drawing a **floral diagram** and a **longitudinal section** through the flower.

A floral diagram is a ground plan of the flower, showing the numbers and arrangement of the floral leaves (Fig. 109). Begin by making with compasses five concentric circles in pencil. Cut the **axis** (not the flower-stalk, but the stem from which the flower-stalk arises) across at the level of the flower, and insert its cut end as a dot above the outermost circle. The side of the flower next the axis is **posterior,** and that next the bract, i.e. furthest away from the axis, **anterior.** Next put in the bract and bracteoles as in Fig. 109 A. We now come to the flower itself, which is between the axis and the bract. First mark the positions of the five sepals on the outermost circle, taking care to draw the posterior sepal in its right position nearest the axis, and to show how the sepals overlap one another. Next draw the five petals (with their nectaries) on the second circle, noticing that they alternate with the sepals. Now put in the stamens. Many flowers have two circles or whorls of stamens, in which case one whorl would be inserted on the third and the other on the fourth circle (cf. Fig. 117). But the numerous stamens of

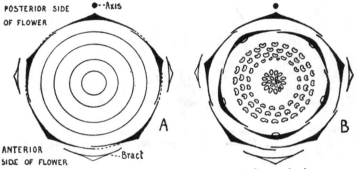

Fig. 109. FLORAL DIAGRAM OF BUTTERCUP. A, how to begin. B, completed diagram.

the Buttercup are arranged spirally, and not in whorls, so you can draw them between the two circles. Complete the diagram by drawing the numerous carpels within the innermost circle (Fig. 109 B).

A longitudinal section. Cut the flower vertically, with a very sharp knife, into two exactly similar halves. Then draw the parts which have been cut through in the way shown in Fig. 110 B. If you show *only the cut surfaces*, your drawing will represent a longitudinal section of the flower. Occasionally it is useful to fill in the parts at the back, in which case the drawing will be that of a half-flower (Fig. 111). Even if this is done, you should start by drawing the longitudinal section first. As a rule, however, it is far better to draw the section only, and not the half-

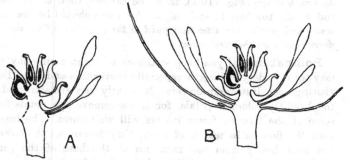

Fig. 110. Longitudinal section through Flower of Creeping Buttercup. A, how to begin. B, completed. (× 4½.)

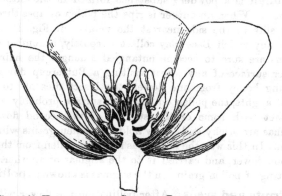

Fig. 111. Half-flower of Creeping Buttercup. (July, × 4½.)

flower, because the reason for drawing the section at all is to show correctly how the floral leaves are inserted on the receptacle. If you compare Figs. 110 B and 111 you will see that the former shows the essential points more clearly.

In drawing a longitudinal section through a flower, *always begin in the centre*, otherwise the important central parts are liable to be crowded. First draw the floral receptacle, showing its exact shape (Fig. 110 A), then the carpels, then the stamens, and finally the corolla and calyx. All parts should be drawn to scale, and *particular attention paid to the insertion of the various floral leaves on the receptacle.*

Pollination. If you keep watch on a bright, sunny day, you may see a variety of insects, especially Hover-flies and small Bees, visiting open Buttercup flowers. Evidently they are attracted by the brightly coloured petals, for if we remove the petals from some of the flowers, fewer insects will visit them. The insects visit the flowers in search of food ; they do not eat the flowers, but suck honey from the nectaries at the base of the petals (Fig. 108 *n*). Bees also collect pollen as food for their young.

Pollen is a powdery substance formed in the anthers of the stamens. When an anther is ripe the pollen escapes through two long slits in the side nearest the petals (cf. Fig. 13 B). Apart from any which Bees may collect purposely, a number of pollen grains are sure to become entangled amongst the hairs on the under surface of any insect visiting a Buttercup flower. After sipping honey from one flower, the insect flies off to another. As it alights, the pollen on its under surface probably comes into contact with some of the stigmas of the second flower. The stigmas are sticky and firmly hold any pollen grains which touch them. In this way the insect has removed pollen from the anthers of one flower, and carried it to the stigmas of another. The depositing of pollen grains on the stigmas is known as **pollination.**

Fruits and seeds. After pollination the flower withers, the sepals, petals and stamens drop off, but the carpels remain (Fig. 112 A). These carpels grow considerably larger and become hard, dry fruits (Fig. 112 B) which finally fall from the receptacle. Each fruit contains a single seed, which has developed from the ovule of the carpel (Fig. 112 c). Unlike the flowers of the

Dame's Violet and Broad Bean (Chap. II) one Buttercup flower gives rise to a number of separate fruits, each of which somewhat resembles a seed in appearance.

We must now enquire whether pollination has anything to do with the formation of seeds.

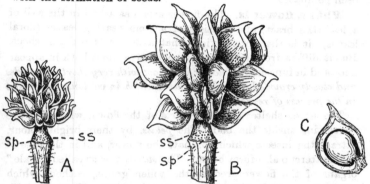

Fig. 112. CREEPING BUTTERCUP. A, carpels after fall of stamens, etc. B, ripe fruits, drawn on same scale as A. C, a fruit cut open showing single seed. *ss*, scars of stamens; *sp*, scars of sepals and petals. (Aug. × 4½.)

EXP. 31. Open very carefully several flower buds of the Buttercup and without injuring the carpels, remove the petals and stamens with a pair of fine forceps. Then cover the flowers with little bags of fine muslin, so that no insects can reach them. In consequence, no pollen can find its way to the stigmas, and you will find that these flowers will produce no ripe fruits or seeds. But the failure to produce seeds might be due to the flower being injured by removing the petals and stamens, and not to the prevention of pollination. To settle this point you should try a control experiment.

Pollinate some of the flowers about two days after removing the stamens, with pollen taken from another flower on a clean, soft brush; afterwards replacing the muslin bags. In this case normal fruits and seeds are formed, so we are justified in concluding that before seeds can be produced the flower must be pollinated.

Fertilization. Pollination, then, is a most important process in the life-history of a Flowering Plant. In this little book we cannot deal with the question of why pollination is necessary, more than to say that botanists have found that part of the living matter of the pollen grain joins or fuses with part of the living matter of the ovule. This fusion of living matter is called **fertilization.** No seed can be formed until fertilization has taken place, and there can be no fertilization unless the flower has first been pollinated.

What a flower is. The flower arises as a bud in the axil of a leaf (the bract) and consists of a stem bearing leaves (floral leaves); it is therefore obvious that a flower is really a shoot. But it differs from an ordinary vegetative shoot both in appearance and in function. *A flower is a shoot with very short internodes and closely crowded specialized leaves which in various ways assist in the process of reproduction.*

The *sepals* shelter the other parts of the flower while they are developing inside the bud. The *petals*, by their bright colour, attract the insects which pollinate the flower, and in the case of the Buttercup also form honey. The *stamens* or so-called "male" organs of the flower produce the pollen grains, without which pollination and fertilization cannot take place. The *carpels* or so-called "female" organs form the ovules and shelter them after fertilization, until at last the seeds are ripe.

HINTS FOR PRACTICAL WORK.

1. Examine some Buttercup inflorescences which show flower buds, flowers and fruits in different stages. Several kinds of Buttercup grow wild in this country; if the Creeping Buttercup is not in flower, any other kind will do.

2. Compare the flowers of any kinds of Buttercup you can find, writing down the characters in which they resemble one another, and those in which they differ. In what habitats did you find the different kinds? Are their vegetative parts alike or different?

3. Construct floral diagrams and longitudinal sections for yourself, using actual flowers, not copying the figures in the book. It is important that these drawings should be on a large scale, considerably larger than there is room for in this little book.

4. In cutting a flower for your longitudinal section, take care to pass through the exact centre, and through both the anterior and posterior sides of the flower. The best way is to start with the stalk and cut upwards.

5. In Exp. 31, the selected flowers must of course be left on the plant. Support the muslin bags with sticks, to prevent injury to the flower-stalks.

CHAPTER XXIII

OTHER FLOWERS

The floral leaves. The stamens and carpels are the floral leaves directly concerned in the production of seeds; they are the **essential organs** of the flower. Sepals and petals are **non-essential** organs, for they only indirectly assist in seed formation.

Most flowers contain all four kinds of floral leaves, but in some flowers one or more may be absent. For instance, the Hazel (Fig.

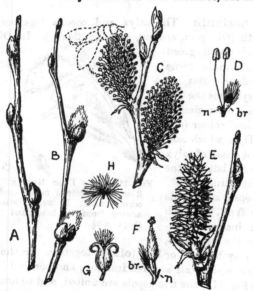

Fig. 113. WILLOW. A, winter twig with catkin buds. B, ditto, hairy catkins appearing. C, male catkins. D, a male flower. E, female catkin. F, a female flower. G, a bursting fruit. H, a plumed seed. *br*, bract; *n*, nectary. (A, B, March, C—F, Apr., G, H, May. A, B, C, E ×3/4; D, F ×3; G, H ×1½.)

121), Birch and Willow (Fig. 113) have two kinds of flowers, either "male" with stamens, or "female" flowers with carpels. In the Hazel and Birch both flowers are found on the same tree, but in the Willow they are on different trees. Again, the flowers of the Ash and Willow have neither calyx nor corolla, the non-essential organs being entirely absent. But in no plant do we find normal flowers with both essential organs (stamens and carpels) missing; they are absent from the double flowers of some cultivated plants, but such flowers are abnormal and do not produce seeds.

The perianth. The calyx and corolla together form the **perianth** (Gk. *peri*, around, *anthos*, a flower). In Dicotyledons the calyx is usually green and the corolla some other colour. But in Monocotyledons the two whorls are often similar in form and colour (e.g. Bluebell), in which case we do not speak of calyx and corolla, but call the whole the perianth.

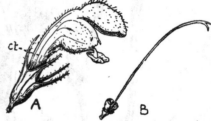

Fig. 114. RED DEAD NETTLE. A, irregular flower with joined sepals and petals; *ct*, corolla tube. B, ovary with long style arising from its base, and two stigmas. (Sept. × 2½.)

The sepals and petals may be **free,** i.e. each may be inserted separately on the receptacle, or they may be more or less **joined** together. In the Buttercup the sepals and petals are free from one another, but in the Dead Nettle (Fig. 114) the five sepals are united, and so are the petals. Many flowers have joined petals, the lower part of the corolla usually forming a more or less definite tube (Fig. 114 *ct*). When the parts are joined you can tell how many sepals or petals there are by counting the number of teeth of the calyx or lobes of the corolla.

Stamens. If the stamens are numerous they are often arranged spirally on the receptacle (Buttercup), but if few, they
are usually in one or two circles or whorls, each whorl having as
many stamens as there are petals in the flower. In this case the
parts of successive whorls generally alternate regularly with each
other, the petals with the sepals, and so on (Fig. 117 A). Stamens
are less often joined to each other than are sepals and petals. In
the Bean family (p. 188) all, or all but one, of the stamens are
joined by their filaments (Figs. 15 s, 156 c, D), while in the Daisy
family (p. 192) the filaments are free, but the anthers joined
(Fig. 125). In flowers with united petals the stamens are usually
joined to the corolla tube (Fig. 123).

Carpels. Most flowers have from 2 to 5 carpels, arranged in
a circle, though the number may vary from one (Sweet Pea, Fig.
156, and Plum) up to many (Butter-
cup, Fig. 112). If the carpels are
free, the flower is **apocarpous**
(Gk. *apo*, away from, asunder,
karpos, fruit), but if the carpels
are joined, the flower is **syncarpous** (Gk. *syn*, together with). In
an apocarpous flower each separate
carpel has its own ovary, style and
stigma, but in a syncarpous flower
there is one **compound ovary**.
The extent to which the carpels
are joined varies in different flowers
(Fig. 115). For instance, in the

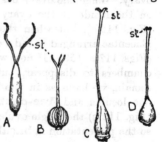

Fig. 115. Compound Ovaries of
Alpine Pink (A), Nasturtium
(B), Foxglove (C), Snapdragon
(D). *st*, stigmas. (Sept. A, B
× 1½; C, D × 1.)

Pink the ovaries alone are joined, the styles and stigmas being
free; in Nasturtium the stigmas and tips of the styles are free;
in the Foxglove only the stigmas are free, while in the Snapdragon the union of the carpels is practically complete.

When the carpels are joined we can generally tell how many
there are, by counting the number of styles or stigmas (Fig. 115).

An even surer way is by cutting the ovary across and examining the cut surface with a lens. Sometimes the ovary is divided into separate chambers or **loculi** (Lat. *loculus*, a little compartment), e.g. Bluebell, Figwort, etc. (Figs. 117, 159); the number of chambers is usually the same as the number of carpels. In other flowers, e.g. Sweet Pea and Violet, the ovary has only a single loculus. In these cases the number of places at which the ovules are joined to the ovary wall gives the number of carpels. You will see at once that the ovary of the Sweet Pea (Fig. 156 c) consists of a single carpel, while that of the Violet (Fig. 142 G) is made up of three.

Placentation. The ovules are borne on the edges of the carpels, and the way in which these ovule-bearing edges (**placentae**) are arranged in the ovary is the **placentation** of the ovary. When the ovary has a single chamber with the placentae on the inside of the ovary wall, the placentation is **parietal** (Fig. 142 G), but when there are two or more loculi, with the placentae arranged round a central axis, the placentation is **axile** (Figs. 117 A, 159). If now we imagine the partitions between the chambers to disappear, but the central axis and placentae to remain, we have, as in the Primrose (Fig. 123 D), an ovary with one loculus and **free-central** placentation. In the Buttercup (Fig. 112 c) the single ovule is inserted at the base of the ovary, so the placentation is **basal.**

Carpels are leaves. When a flower bud is developing, the sepals (which are lowest on the stem) appear first and the carpels last. In this and in other respects, though they have no buds in their axils, all four kinds of organs of the flower agree with vegetative leaves (p. 66), so we can call them all "**floral leaves.**" The sepals and petals are more or less leaf-like in appearance, the stamens usually less so, while the carpels, especially when joined together, often do not look like leaves at all. But if we split open a fruit of the Larkspur, it is easy to see its leaf-like nature (Fig. 116). The leaf-blade has a network of veins with a distinct midrib, and a strong vein running along each edge (**placenta**) of the leaf. The marginal veins supply the ovules with food and water. The midrib of the carpellary leaf is prolonged to form the style, with the stigma at its extreme tip.

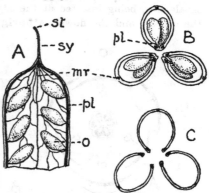

At a very early stage of development the carpel curves or bends until its edges meet, and a closed cavity, the ovary, is formed. The joined edges of the leaf (placentae) are now enclosed within the ovary (Fig. 116 B). Later, when the seeds are ripe, the edges of the carpels again separate from one another (Fig. 116 c). The

Fig. 116. LARKSPUR. A, upper part of carpel, opened out to show ovules. B, three carpels cut across. C, ditto, after ripening of fruit. *mr*, midrib; *sy*, style; *st*, stigma; *pl*, placenta; *o*, ovule. (Sept. ×2.)

carpels of a syncarpous flower develop in a similar way, but soon become joined together. If they join *after* the bending described above, the result is a compound ovary with several chambers and axile placentation (Bluebell, Fig. 117). If, however, the carpels unite by their adjacent edges *first*, then they cannot bend further and we have a compound ovary with one chamber and parietal placentation (Violet, Fig. 142 G).

How the ovules are sheltered. Within the closed ovary, whether simple or compound, the ovules are protected from drying and other dangers, until they have developed into ripe seeds.

Floral receptacle. In the Buttercup, Bluebell and Primrose the receptacle is more or less conical, the sepals, petals and stamens being inserted on it at a lower level than, i.e. *below*, the carpels or "female" organs (Figs. 110, 117 B, 123). Such flowers are said to be **hypogynous** (Gk. *hypo*, beneath, *gyne*, the female). In other cases the receptacle is cup- or basin-shaped (Fig. 118). The sepals, etc., being inserted at the edge of the basin, are *around* the carpels, and the flower is **perigynous** (Gk. *peri*, around).

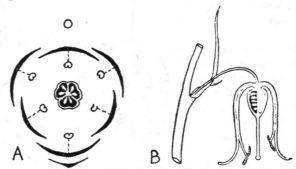

Fig. 117. HYPOGYNOUS FLOWER OF BLUEBELL. A, floral diagram. B, longitudinal section. (May, × 1½.)

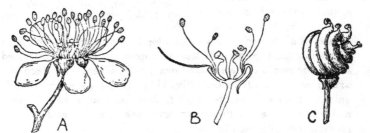

Fig. 118. PERIGYNOUS FLOWER OF MEADOW SWEET. A, flower. B, longitudinal section. C, twisted fruits. (Aug. × 4.)

Finally, the cup-shaped receptacle may be actually joined to the wall of the ovary, so that the other floral leaves are inserted *above* the carpels. Such flowers are **epigynous** (Gk. *epi*, upon), e.g. Hogweed, Iris (Figs. 119, 130).

In hypogynous and perigynous flowers the ovaries are **superior,** but in epigynous flowers they are said to be **inferior.** In some perigynous, but more particularly in epigynous flowers, the developing seeds are protected by the receptacle as well as by the ovary wall.

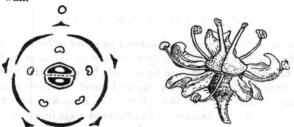

Fig. 119. EPIGYNOUS FLOWER OF HOGWEED, and its floral diagram. (Flower, Aug. × 6.)

HINTS FOR PRACTICAL WORK.

1. Try to find out for yourself which types of flowers are most common. Do you find more kinds of flowers with free or with joined petals? Which is the more frequent, apocarpy or syncarpy; hypogyny, perigyny or epigyny?

2. Practise drawing ground plans (floral diagrams) of, and longitudinal sections through, various flowers (see Chap. XXII). If a flower has joined sepals or petals, or epipetalous stamens, you should show this in the floral diagram (cf. Figs. 159, 117 A). In making floral diagrams, a good plan for beginners is to make five concentric circles on a piece of sheet cork, and to lay out the actual parts of the flower on the circles. As you dissect the flower, pin each part in turn in its correct position. Laying out the flower first in this way will help you to make the ground plan properly, but such points as the overlapping of petals, etc., can of course only be made out from an undissected flower. The cork can be kept and used for other flowers.

3. The most difficult part of the flower to understand is the central or "female" part. Examine carpels of Larkspur, Columbine or Marsh Marigold at various stages of development, and try to understand how each carpellary leaf is folded to form an ovary. Then compare the apocarpous condition of these flowers with the syncarpous condition found in the Bluebell and Violet.

4. Make some good-sized plasticine models of single open (i.e. not folded) carpels. By folding and joining these in various ways you can build up models of different kinds of syncarpous ovaries. Then cut your models across to see the placentation.

5. In constructing a floral diagram, pay particular attention to the placentation of the ovary; this is often more easily studied by using half-grown fruits, in which the ovary is larger than in the flower itself.

CHAPTER XXIV

CROSS- AND SELF-POLLINATION

Nature often attains the same end in many different ways, and nowhere is this more strikingly seen than in the case of flowers. The business of flowers is to form seeds by which the plant may be reproduced. Before seeds can be formed the flowers must be pollinated, and much of the diversity of form and structure we find in flowers has to do with the different ways in which the very important preliminary process of pollination is effected.

Cross- and self-pollination. If pollen from one flower is placed on a stigma of another flower of the same kind, the second flower is said to be **cross-pollinated;** but if pollen is merely transferred from an anther to a stigma of the same flower, the flower is **self-pollinated.** Pollen may be carried from one flower to another in various ways, the two most frequent methods being by the aid of either wind or insects.

Cross-pollination by wind. Most grasses are cross-pollinated by means of wind, though a few are self-pollinated. A grass flower (Fig. 120) is inconspicuous, for though there are small greenish bracts, there is no perianth. The flower consists mainly of three stamens, and an ovary with two large feathery stigmas. The dry, powdery pollen is shed from the anthers

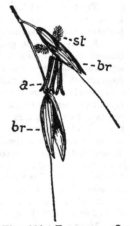

Fig. 120. FLOWER OF OAT GRASS. *a,* anthers; *st,* feathery stigmas; *br,* bracts. (July, × 2.)

only in dry weather, and the anthers, loosely attached to their filaments, hang freely out of the flower (Fig. 120). When a breeze is blowing, the pollen is shaken out of the anthers and carried away by the wind. If some of this floating pollen chances to reach the projecting, feathery stigmas of another flower, pollination is effected. Self-pollination is often prevented in grasses by the stigmas becoming ripe before the anthers.

The Great Plantain (Fig. 133) and the Hazel are also wind-pollinated. The "male" flowers of the Hazel are arranged in drooping, yellowish catkins ("lambs' tails"), and are far more

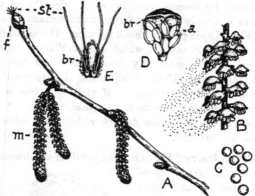

Fig. 121. HAZEL. A, twig with male catkins (*m*) and female flowers (*f*). B, part of male catkin shedding pollen. C, pollen grains, magnified. D, male flowers. E, two female flowers. *a*, anthers; *st*, stigmas; *br*, bracts. (Mar. A × 3/4; B × 2; D, E × 4.)

numerous than the inconspicuous "female" flowers, which occur in clusters rather resembling winter buds (Fig. 121). When the Hazel is flowering the crimson stigmas project and readily catch some of the pollen carried from the catkins as they are swayed by the wind.

Many of our trees have inconspicuous wind-pollinated flowers. Some, e.g. Elm (Fig. 100), Poplar, Alder, Hazel, flower early in the year, before the leaves appear; this no doubt makes it easier for wind-borne pollen to reach the stigmas. In others such as Birch and Oak, flowers and leaves unfold at about the same time, while in the Beech the leaves appear first. Some trees, e.g. Willow, Sycamore and Horse Chestnut, have insect-pollinated flowers. In wind-pollinated flowers such vast quantities of pollen are produced, that though most of it is lost, some will probably reach the stigmas of other flowers.

Cross-pollination by insects. Far more kinds of flowers are pollinated by insects than by wind. The Buttercup (Chap. XXII) has a simple **pollination mechanism;** that of the Willow (Fig. 113) is perhaps even simpler. The small flowers are clustered together in catkins, which in most Willows appear before the leaves. Insects, especially Bees, crawl over the catkins in search of honey (Fig. 113 c). While doing this, they either receive pollen on their under surfaces, or, if visiting a tree with female flowers, transfer this pollen to the stigmas.

We must now study a few flowers which have more elaborate pollination mechanisms.

The Cross-leaved Heath. The petals of the drooping corolla are joined, the narrow opening of the corolla tube being occupied by the stigma (Fig. 122 A). The pollen escapes through pores at the tip of the anther, and each anther has two long horns. The nectary is underneath the ovary. When a Bee visits the flower, its head touches the stigma first, and if the insect has come from another Heath flower, pollen will be transferred to the sticky stigma.

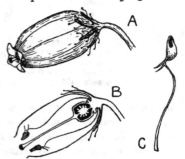

Fig. 122. CROSS-LEAVED HEATH. A, flower. B, longitudinal section. C, a stamen. (July, A, B × 3; C × 4½.)

In probing for honey, the Bee's tongue is sure to touch some of the anther horns, and to shake pollen out of the anther. The pollen falls on the insect's head and is carried away to another flower. Only insects with tongues as long as the corolla tube can reach the honey and pollinate the flower.

The Primrose has two kinds of flowers (Fig. 123). In a long-styled or "pin-eyed" flower the stigma is at the top of the corolla tube and the stamens half-way down, while in a short-styled or "thrum-eyed" flower the positions of the stigma and stamens are reversed. When a Bee visits a Primrose flower for honey (underneath the ovary), it receives pollen either on its head or on its tongue, according to whether the flower is thrum- or

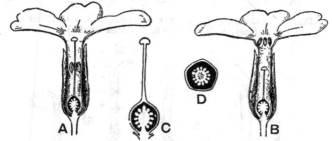

Fig. 123.   PRIMROSE.   Longitudinal sections through (A) pin-eyed and (B) thrum-eyed flowers, and (C) ovary, etc.   D, cross section of ovary, showing free-central placentation.   (Mar. A, B × 1¼; C, D × 2½.)

pin-eyed. In visiting other flowers the pollen from a short-styled flower will naturally be carried to the stigma of a long-styled flower, and the pollen of a long-styled to the stigma of a short-styled flower. Cross-pollination is thus almost sure to take place.

In the flower of the **Germander Speedwell** the two anthers are so far away from the stigma (Fig. 124 A) that at first sight it is difficult to imagine how pollen can reach the stigma. The Speedwell is commonly pollinated by Hover-flies, which alight on the flower in such a way as first to touch the stigma with the under surface of the body. To support itself the insect grasps the stamens, and in so doing draws them under its body (Fig. 124 B), thus receiving a fresh load of pollen which it carries to the stigma of another flower. The stamens move very easily, owing to the thin, hinge-like base of the filament (Fig. 124 c).

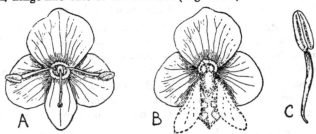

Fig. 124. FLOWERS OF GERMANDER SPEEDWELL. B, with Hover-fly. (July, A, B × 3; C × 6. Fly after Knuth.)

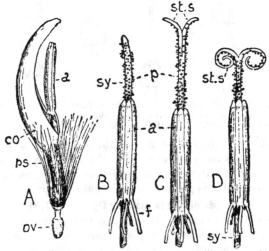

Fig. 125. DANDELION. A, young flower. B—D, stages in pollination (stamens, styles and stigmas only shown). *co*, corolla; *ps*, pappus (= calyx); *ov*, inferior ovary; *sy*, style; *st. s*, stigmatic surfaces; *f*. filament; *a*, anther tube; *p*, pollen. (Aug. A × 4; B—D × 6.)

The so-called "flower" of the **Dandelion** is really an inflorescence (p. 167) containing many small flowers, one of which is shown in Fig. 125 A. The five stamens are joined to the corolla, their filaments being free, but the anthers united to form a narrow tube round the style (Fig. 125 *a*). The pollen is shed into this anther tube, and is swept out by the hairy style as it grows up through the tube (Fig. 125 A, B). If at this stage an insect crawls over the flowers it carries away some of the pollen. Now the two stigmas separate (Fig. 125 c), and may be cross-pollinated by an insect carrying pollen from another plant. You might think that self-pollination would occur as the style brushes the pollen out of the tube. This is not the case, for pollen must be placed on the *upper* surface (= **stigmatic surface**) of the stigmas, and at first these are tightly pressed together (Fig. 125 B). It is only when the stigmas separate and expose the stigmatic surfaces that pollination can occur. But the flower may be self-pollinated after all, for the stigmas keep on growing till they curl round and touch their own pollen (Fig. 125 D). The pollination mechanism of the Dandelion is very effective, for if the flower is not cross-pollinated by an insect, self-pollination is almost sure to occur. Most plants cannot produce ripe fruits or seeds unless the flower has first been pollinated (cf. Exp. 31, p. 143), but the Dandelion and a few other plants may form seeds even without pollination.

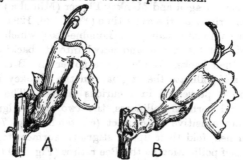

Fig. 126. FLOWERS OF WOOD SAGE. A, stamens ripe. B, an older flower, stigmas ripe and stamens withering. (July, ×2½.)

most flowers contain both stamens and stigmas, these organs do not as a rule ripen at the same time. Such flowers are usually not self-pollinated. In the Wood Sage, for example, the stamens are ready first, taking up a position in which they are sure to dust with pollen the back of a Bee visiting the flower for honey (Fig. 126 A). Later, when the pollen has been shed, the stamens move backwards out of the way. At the same time the style bends forward, bringing the stigmas into the correct position for pollination (Fig. 126 B).

In the Figwort, which is pollinated by Wasps, the stigma ripens before the stamens. After about two days the stigma withers,

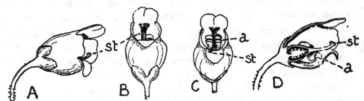

Fig. 127. FIGWORT. A, B, stigma (*st*) ripe first. C, stigma withered, anthers (*a*) ripe. D, flower A cut in half. (July, × 2¼.)

while the bent stamens straighten out and occupy its position (Fig. 127).

In the Snapdragon and Monkey Flower (Mimulus) the anthers and stigma ripen at the same time (Figs. 128, 129). The Snapdragon is pollinated mainly by Humble-bees, which force apart the two lips of the flower and crawl into it, backing out with pollen on their backs. While in the flower the Bees may either cross- or self-pollinate the stigma. In the Monkey Flower self-pollination is prevented in a curious manner. The insect first touches, and perhaps pollinates, the two large stigmas. These stigmas are sensitive to contact (cf. tendrils, p. 74), and on being touched, fold their upper, stigmatic surfaces together; this prevents self-pollination as the Bee retires (Fig. 129 B, C).

Self-pollination is prevented in the Iris in a somewhat similar

way. There are three large petal-like styles, the stigmatic surface
being on the upper side of a little flap near the tip of each style.

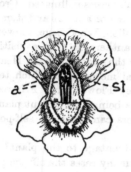

Fig. 128. FLOWER OF SNAP-
DRAGON WITH LOWER LIP PULLED
DOWN. *st*, stigma lying beween
the four anthers (*a*). (July,
× 3/4.)

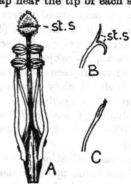

Fig. 129. MONKEY FLOWER. A,
sensitive stigmas projecting beyond anthers. B, section through
open stigmas. C, ditto, stigmas
closed. *st.s*, stigmatic surfaces.
(July, × 1½.)

There are three stamens, one
beneath each style (Fig. 130).
A Humble-bee alights on one
of the sepals and most likely
bends back and pollinates the
stigmatic flap. It then creeps
underneath the stamen to the
honey at the base of the perianth, receiving a new load of
pollen. By this time the flap
has sprung upwards again,
which prevents self-pollination
as the Bee backs out of the
flower.

**The advantages of
cross-pollination.** Many

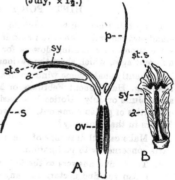

Fig. 130. FLOWER OF IRIS. A, longitudinal section. *s*, sepal; *p*, petal;
a, anther; *ov*, inferior ovary; *sy*,
petal-like style; *st.s*, stigmatic surface. B, a style and stamen. (July,
× 1/2.)

plants are at least occasionally self-pollinated, and some, such as Groundsel and Sweet Pea, are nearly always self-pollinated. But the great majority of flowers are generally cross-pollinated. Cross-pollination is favoured by: (1) the ripening of anthers and stigmas at different times, e.g. Buttercup, Dandelion, Wood Sage, Figwort; (2) the stigma projecting beyond the anthers, so that a visiting insect naturally comes into contact with the stigma first, e.g. Heath, Speedwell, Monkey Flower; (3) special mechanisms which tend to prevent self-pollination, e.g. Monkey Flower, Iris, Primrose; and finally, (4) male and female flowers being on different plants, e.g. Willow. In the last case the flowers can never be self-pollinated.

Is cross-pollination of any special advantage to the plant? It has been found by experiment that in many cases the offspring of cross-pollinated plants are taller, stronger, and more numerous than the offspring of self-pollinated plants. We may conclude then, that though self-pollination is not injurious, yet for many plants cross-pollination is decidedly beneficial.

HINTS FOR PRACTICAL WORK.

1. By this time you have probably found that it is much easier to understand and remember what you read if you examine actual specimens at the same time. In collecting flowers for this purpose try to find specimens of different ages, which show the various stages described.

2. Gently shake some wind pollinated plants when the stamens are just ripe, e.g. Hazel about February or March, Pine about May, many grasses about June or July. Notice how easily the stamens move, and what great quantities of pollen come out. Can you shake pollen out of insect-pollinated flowers in the same way?

3. Make careful drawings of the flowers you examine, and especially the parts concerned with pollination.

4. Try to find answers to the following questions:

(a) Can you find nectaries in any wind-pollinated flowers?

(b) Do pin- and thrum-eyed Primroses grow on the same or on different plants?

(c) In the Iris, Buttercup, Heath, etc., which part of the insect's body receives the pollen? On which side of the anther is the pollen shed—on the side towards the insect or on the side away from it?

CHAPTER XXV

POLLINATION CONTINUED. INFLORESCENCES

How flowers attract insects. Insects are useful to many flowers as agents of cross-pollination. They carry out this process unintentionally, and probably unconsciously, while in quest of the food (either honey or pollen) with which the flower provides them. Insects are attracted to flowers in the first place by means of the senses of smell and sight. Many flowers have distinctive odours which indicate to the insect, often before it can see the flowers themselves, the direction in which to fly. As it comes nearer, the bright colours of the petals complete the attraction, and the insect is guided by sight as well as by smell. Many flowers too have lines or spots of a colour different from that of the rest of the corolla. Some botanists think that these help intelligent insects to find the honey. The lines on the corolla of the Speedwell, and the yellow ring round the centre of the flower of Forget-me-not, are examples of these so-called "honey-guides."

Nectaries. The nectaries are often just below the ovary,

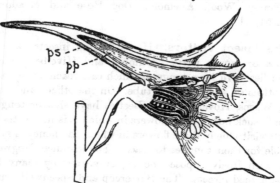

Fig. 131. IRREGULAR FLOWER OF LARKSPUR CUT IN HALF LONGITUDINALLY. *ps*, posterior spurred sepal; *pp*, spurred petal. (July, ×2.)

e.g. Primrose and Heath (Fig. 122). In Dandelion they are above the ovary, and in the Dame's Violet (Fig. 13 n) at the base of the two short stamens. The Buttercup has a nectary at the bottom of each petal (Fig. 108 n), while in the Larkspur and Monkshood, which belong to the Buttercup family, the honey-forming petals are greatly modified. The two posterior petals of the Larkspur (Fig. 131) form long, honey-containing "spurs," enclosed within a third spur, the posterior sepal. In Monkshood the sepals are coloured and petal-like, two of the petals forming stalked nectaries (Fig. 132).

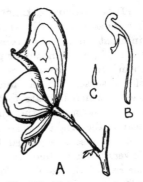

Fig. 132. MONKSHOOD. A, flower with petal-like calyx. B, nectary. C, a rudimentary petal. (Aug. A, B × 1; C × 4.)

Some flowers do not form honey, and offer only pollen to their visitors. Examples of such **pollen flowers** are Poppy, Wood Anemone, Dog Rose and Meadow Sweet (Figs. 54, 118).

The insect visitors. Honey-feeding insects have sucking tongues or mouth parts. Humble- and Hive-bees, Moths and Butterflies have long tongues, which can reach honey hidden at the bottom of a long corolla tube. On the other hand, some wild Bees, Wasps, Hover-flies and Beetles have shorter tongues, and can only suck honey from flowers in which it is nearer the surface. Insects visit chiefly those flowers in which the honey is at a depth suitable for their tongues to reach. For instance, Hogweed (Fig. 119), with freely exposed honey, is visited by many kinds of short-tongued insects. The Buttercup and Speedwell, with easily reached honey, are commonly visited and pollinated by short-tongued Hover-flies, though other visitors are not infrequent. The Figwort is a Wasp-flower, and Clover, Heath and Wood Sage are Bee-flowers. The Snapdragon is visited especially by Humble-bees, which are heavy enough to force apart the two lips of the flower, while Honeysuckle, with its very long corolla tube, is pollinated chiefly at night by large moths.

Bees. Many flowers conceal their honey at the bottom of long tubes or spurs ; this tends to exclude short-tongued, and to en-courage long-tongued visitors such as Bees. Bees are the most important agents of pollination. They feed themselves and their young entirely on honey or pollen, to obtain which they must visit enormous numbers of flowers. Bees too, more than other insects, tend to keep to one kind of flower during a particular flight. This enables them to work more quickly, and at the same time benefits the flowers, which are more likely to receive the right kind of pollen. I once watched a Hive-bee working over a Daisy-covered lawn. Not a single Daisy was visited, but only the less conspicuous flowers of White Clover. Bees often work very quickly when collecting honey. One sunny afternoon in May in my garden a Hive-bee visited 115 Forget-me-not flowers in 3 minutes. On a June evening, no doubt after a long day's work, another Bee visited 30 flowers of the Broad Bean in 5 minutes. These observations will give some idea of the rate at which Bees work ; you should make others for yourself.

Pollination mechanisms. We saw in the last chapter that different parts of the visiting insects may be dusted with pollen ; in some flowers it is the head of the insect, and in others either the back, under surface or tongue. But whatever the part may be, the really important thing is that *the stigma and the anthers should both touch the same part of the insect's body*, otherwise the stigma will not be pollinated. How this may be effected without at the same time causing self-pollination you will have learned from the flowers described.

Flowers such as the Buttercup and Heath (Figs. 111, 122) are said to be **regular,** which means that if the flower is cut vertically in two, passing through the middle of any one of the petals, the two halves will be similar. The Dead Nettle, Speedwell and Larkspur (Figs. 114, 124, 131), on the other hand, have **irregular** flowers, that is, they can only be cut vertically into similar halves if the cut passes through both the anterior and posterior sides of the flower. As a rule insects can only obtain honey from an irregular flower if they alight in one particular way ; the stamens and stigma being placed so as to touch an insect entering in that position. It is usually the cleverer insects such as Bees which visit irregular flowers.

The corolla often serves as an **alighting platform,** e.g. the spreading part of the corolla of the Dame's Violet and Primrose (Figs. 12, 123), and the lower lip of the Wood Sage and Snapdragon (Figs. 126, 128).

You should study very carefully a few of the numberless floral mechanisms. When you know the structure of a flower really thoroughly, then watch the insects at work, always trying as you do so to " see the flower as the Bee sees it."

Inflorescences.

Flowers are generally borne towards the tip of the main stem or of its branches. This position aids the pollination of both wind- and insect-pollinated flowers, for the top of a plant is more exposed to wind than the lower parts, and is also more easily seen by flying insects. In some plants the flowers are **solitary,** i.e. borne singly, in either a terminal (e.g. Tulip) or an axillary (e.g. Creeping Jenny) position. Far more frequently, however, the flowers are clustered together into branched **inflorescences,** which may be very conspicuous, so increasing the distance from which the flowers can be seen by insects. The flowers in an inflorescence normally arise in the axils of leaves called **bracts** (p. 65), though in many inflorescences the bracts are absent (Fig. 12). Like vegetative shoots, inflorescences may branch either monopodially or sympodially (cf. Chap. XIX). We will first study some **monopodial,** or, as they are often called, **racemose** inflorescences.

The raceme has an elongated stem of more or less indefinite growth, which bears a number of *stalked* flowers; the flower-stalks are shorter than the main stem. The flowers are developed in the same order as the leaves on a vegetative shoot, i.e. the oldest below and the youngest above. Examples of racemes are Dame's Violet, which has no bracts (Fig. 12), and the Currant and Bluebell, in which bracts are present.

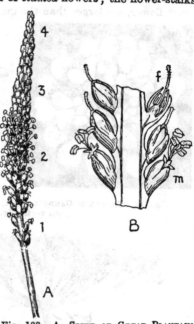

The spike is similar to the raceme, but the flowers are **sessile,** that is without stalks, e.g. Great Plantain (Fig. 133). The **catkin** is a spike containing only male or female flowers, e.g. Willow (Fig. 113). The male catkins of wind-pollinated plants such as the Birch, Alder and Hazel (Fig. 121) usually hang downwards, in which position they are easily swayed by the wind.

Fig. 133. A, Spike of Great Plantain. The flowers open from below upwards, and the stigmas are ripe before the stamens; 1, young fruits; 2, flowers with ripe stamens; 3, flowers with ripe stigmas; 4, unopened buds. B, part of spike cut vertically; *f*, "female"; *m*, "male" stage. (July, A × 3/4; B × 3.)

The corymb. If the stalks of the lateral flowers of a raceme were to grow as fast as the main axis, we should have an inflorescence known as a corymb. A corymb is very conspicuous to insects flying overhead, for all the flowers are at very much the same level. In Candytuft the inflorescence is particularly conspicuous because the outer flowers, as well as the outer petals of each flower, are larger than the inner ones (Fig. 134).

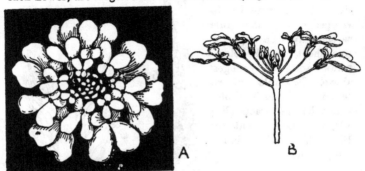

Fig. 134. CORYMB OF CANDYTUFT. A, as seen from above by an insect. B, cut vertically. (July, × 1.)

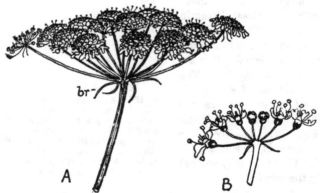

Fig. 135. A, COMPOUND UMBEL OF HOGWEED. B, a simple umbel cut vertically. *br*, bracts. (Aug. A × 3/4; B × 1½.)

The umbel. The Ivy has a **simple umbel** (Lat. *umbella*, a sunshade), which resembles a corymb, except that all the flowers spring from about the same level on the main stem, like the ribs of an umbrella. The Hogweed (Fig. 135) has a **compound umbel,** or "umbel of umbels"; the top of the stem branches like an umbel, each branch bearing a simple umbel at its tip.

The capitulum resembles a simple umbel with sessile flowers borne on the enlarged tip of the stem, e.g. Daisy (Fig. 136) and Dandelion. The whole inflorescence, with its calyx-like ring of bracts (the **involucre**), looks like a single flower.

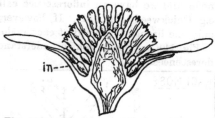

Fig. 136. CAPITULUM OF DAISY. The outer flowers (ray florets) are irregular, the inner (disc florets) regular. *in*, involucre of bracts. (Aug. ×3.)

In the racemose or monopodial inflorescences described above, *the lowest or outermost flowers are the oldest, the youngest flowers being nearest the top or the centre of the inflorescence.*

In **sympodial** or **cymose** inflorescences, each branch usually bears either one or two bracteoles (p. 65), and then

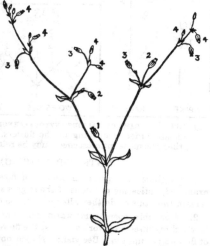

Fig. 137. TWO-SIDED CYME OF CHICKWEED. Numbers indicate the age and order of opening of the flowers. (Aug. ×1/2.)

ends in a flower. Later branches arise in the axils of the bracteoles, and these branches too end in flowers. This is much like the sympodial branching of vegetative shoots (Chap. XIX). If the plant has opposite leaves, two branches may arise at each node, and we have an inflorescence called a **two-sided cyme,** e.g. Chickweed (Fig. 137). If, however, the leaves are alternate, only one branch is formed at each node (**one-sided cyme**).

Fig. 138 may help you to understand the various kinds of inflorescences.

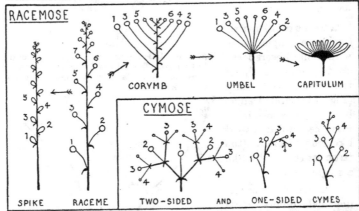

Fig. 138. DIAGRAMS OF VARIOUS INFLORESCENCES. Numbers show the relative age and order of opening of the flowers. The arrows indicate how the other racemose inflorescences may be related to the raceme.

HINTS FOR PRACTICAL WORK.

1. Continue your practical studies of flower pollination. In each flower examined, notice the positions of the stigma and anthers. Do these positions remain the same or do they alter with the age of the flower?

2. If possible, spend some warm, sunny afternoons in spring and summer in watching insects at work amongst the flowers. Make notes of the numbers of flowers per minute a Bee visits. Notice how a Bee (or other insect) enters different kinds of flowers; also the masses of pollen on the hind legs of Bees visiting pollen flowers.

3. Collect and examine different kinds of inflorescences. Selecting typical examples, construct diagrams to show the arrangement of flowers and bracts in the inflorescence, and the order of opening of the flowers.

CHAPTER XXVI

FRUITS AND THE MIGRATIONS OF PLANTS

When once pollination has been effected, floral mechanisms such as those we have been studying are no longer required. The perianth and stamens wither and usually fall, but the ovary remains and gradually enlarges and ripens into the **fruit,** which contains the seeds developed from the fertilized ovules (Chaps. II and XXII).

The number of fruits formed from one flower depends on the number of ovaries in the flower. Most flowers have only one ovary, which gives rise to a single fruit. This may be formed from one carpel (Vetch and Plum, Figs. 141 D, 150 D), or from more than one, e.g. two in Dame's Violet, three in Dog Violet (Figs. 14, 142). In the Buttercup and other apocarpous flowers a single flower produces a number of separate fruits, each consisting of one carpel (Figs. 112, 148, 150 A, 152).

The fruit has two main functions: to protect the seeds while they are developing, and to help to scatter or disperse them when they are ripe.

Seed-dispersal. A plant cannot move about from place to place as an animal can, but spends its life rooted to the spot where the seed germinated. It is true that plants with vegetative means of propagation travel to some extent, but for most plants the only chance of travelling or migrating is during the seed stage. Now the number of seeds produced by even a single plant is enormous. For example, the common Flixweed, an annual herb, usually produces something like 730,000 seeds. It has been estimated that if every one of these seeds survived and came to maturity, the descendants of one plant might cover in three years an area 2000 times as great as the whole land surface of the earth. Now try to imagine what would happen if all these seeds were to fall and germinate close to the parent plant. You can get some idea of the result by performing a simple experiment.

EXP. 32. Fill two similar flower pots with good garden soil. In one pot sow about half a dozen Mustard seeds, leaving plenty of room for each to grow. In the other pot sow Mustard seeds very thickly. Cover the seeds lightly with soil and allow them to grow, watering when necessary.

Fig. 139 shows the result of such an experiment. The seeds were sown on July 16, and at first all the seedlings were similar.

But notice the difference three weeks later. B, C and D (Fig. 139) are from the crowded pot; they are weak and partially etiolated, for they have had to compete with each other for the very necessaries of life—light, water, salts—and there was not enough to go round. Few of the seedlings in this pot were as large as B, most resembled C or D. A month later many had died altogether, while others had scarcely grown at all. If you compare these seedlings with A, from the uncrowded pot, you will see not only the ill effects of overcrowding, but also the advantage of lessened competition for the necessaries of life. A good gardener

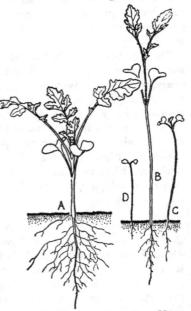

Fig. 139. THREE WEEKS OLD MUSTARD SEEDLINGS. EXP. 32. A, uncrowded. B—D, from a crowded pot. (Aug. × 1/3.)

knows these things, and always tries to grow vigorous, healthy plants. He lessens competition by keeping down weeds and, where necessary, by thinning out or transplanting young seedlings.

But what actually happens in nature? The land surface of the globe is so covered with plants already that suitable vacant spaces are not very abundant. Further, what is suitable for one kind of plant may not be suitable for another; for instance, a Dandelion cannot thrive in a shady wood, or Wood Sorrel on a wind-swept moorland. How do seeds reach really suitable vacant places? To a great extent this is due to chance, for a seed cannot wander about at will until it finds an ideal spot in which to germinate. Every year vast numbers of seeds are produced and scattered far and wide. Most of these seeds perish, either before or after germination, some because they are eaten, and others because they fall on ground which is either unsuitable or already occupied. But a few may chance to fall on favourable ground where they can spring up, grow to maturity, and in turn reproduce their kind.

Agents of dispersal. From what has been said above it follows that seed-dispersal is a very important process. We shall see that the structure or mechanism of the fruit or seed often aids dispersal, but the seeds themselves have no power of self-movement. For this reason seeds can only be transported to a distance by some external agency. Sometimes the wall of the fruit itself is the agent of dispersal, but more commonly the agents are wind, water or animals.

Changes during ripening. As the ovary grows, its soft green wall undergoes various changes, finally becoming the coat or **pericarp** of the ripe fruit. In the so-called **dry fruits,** the pericarp becomes hard, dry and brown, but in **succulent fruits** it becomes soft and fleshy, often with a brightly coloured skin.

Fruits which are dry when ripe.

Separation of seeds. Some fruits have only one seed, but the majority contain a number of seeds; sometimes, as in the Poppy, Foxglove, and especially in Orchids, the number is very large indeed. If the danger of overcrowding (Exp. 32) is to be avoided, it is necessary that these seeds should be separated from one another before germination. Dry fruits which contain a number of seeds split open, or **dehisce,** when ripe, allowing the seeds to escape and to be separated. Such fruits are said to be **dehiscent** (Lat. *dehisco*, I yawn). Fruits which only contain one seed do not open in this way; they are **indehiscent.** In these cases the seeds are already separate, for they are enclosed in different fruits (Fig. 112).

One-seeded dry fruits. The two chief kinds of dry, indehiscent fruits are the **achene,** which is small (e.g. Buttercup, Dandelion, Strawberry, Figs. 112, 145, 152), and the **nut,** which is larger and has a hard, often woody pericarp (e.g. Hazel, Oak, Beech). The fruits of the Elm (Fig. 100 F) and Ash (Fig. 145 A—C) are winged nuts or **samaras.** Although a *ripe* nut only contains one seed, there may be to begin with several ovules in the ovary. Fig. 145 C shows the ovary of the Ash in June; there are four ovules, but one of them, which will become the single seed (Fig. 145 B), is already growing at the expense of the other three.

In most cases it is easy to tell a fruit from a seed, but small one-seeded fruits may be confused with seeds, so it is well to know how to distinguish them from one another. Seeds have *one* coat, the testa, and *one* scar, the hilum (Chap. III), while one-seeded fruits have *two* coats, with *two* scars on the outer coat. The two coats are the pericarp and the testa, and the two scars are where the stigma and flower-stalk respectively were attached to the ovary. Sometimes the withered remains of the stigma itself or of the calyx or other organs are still visible. The case of the one-seeded dry fruits of grasses (e.g. Wheat and Maize, p. 21) is the most difficult one you are likely to come across. The pericarp and testa are joined, so that the grain appears to have only one coat; at the same time the scar of the stigma is particularly small and easily missed.

"Split-fruits." Some fruits which contain several seeds do not dehisce in the ordinary way, but the seeds are separated by the fruit itself splitting into several one-seeded portions. Examples are Sycamore (Fig. 18), Geranium, Hogweed, Garden Nasturtium and Woundwort (Fig. 140). The fruit of Geranium dehisces as well as splits into pieces. In the first four plants named the fruit splits into as many portions as there are carpels, but in the Woundwort the fruit splits into four "**nutlets**," though there are only two carpels.

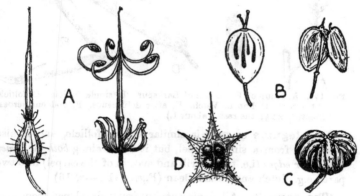

Fig. 140. "SPLIT-FRUITS" of A, wild Geranium. B, Hogweed. C, Nasturtium. D, Woundwort. (Aug. C × 1, the rest × 2.)

Dry fruits with more than one seed. The seeds of indehiscent dry fruits, and of "split-fruits," generally have a thin testa, the seeds being sufficiently protected during dispersal by the pericarp. But the seeds of dehiscent fruits, which escape from the fruit before dispersal, usually have a tough protective testa of their own.

There are three chief kinds of dehiscent dry fruits, the follicle, legume and capsule.

The follicle is a pod-like fruit, consisting of a single carpel, which dehisces when ripe along its *inner edge*. Follicles are common in the Buttercup family, e.g. Larkspur (Figs. 141 A—c, 116 c),

Columbine and Marsh Marigold. Several follicles are usually developed from one flower.

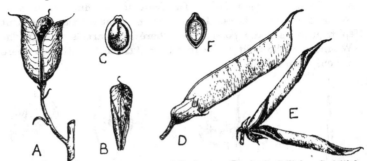

FIG. 141. A, group of FOLLICLES of Larkspur. B, single follicle. C, follicle cut across. D, POD of Vetch. E, after dehiscence. F, pod cut across. (Sept. C × 2½; the rest × about 1.)

The legume or pod is similar to the follicle, and is also developed from a single carpel, but it splits along *both the inner and outer edges* (i.e. the midrib and margin of the carpel) into two parts, e.g. Vetch and Broad Bean (Figs. 141 D—F, 16).

The capsule. All dehiscent dry fruits developed from syncarpous ovaries (and therefore composed of two or more carpels) are called capsules. There are many different kinds of capsules, a few of which are shown in Fig. 142 (also Fig. 14). Capsules dehisce in various ways, e.g. by teeth, as in Campion and Mouseear Chickweed (Fig. 142 A, B); by pores in Poppy (C) and Campanulas; by a lid in the Pimpernel (D) and Plantain, or as in most capsules, by splitting from the top downwards, often by as many splits as there are carpels, e.g. Figwort and Dog Violet (E, F).

HINTS FOR PRACTICAL WORK.

1. In Exp. 32 keep the pots under observation for some months. Make notes from time to time of the appearance, size, etc. of the plants in the two pots.

2. Collect and examine as many of the different kinds of dry fruits mentioned in this chapter as you can find.

8. In order to get an idea of the numbers of seeds formed in dry fruits, count the seeds in a few nearly ripe fruits, e.g. Violet, Wallflower, Foxglove.

4. Make large-scale drawings of a few one-seeded dry fruits, e.g. Buttercup, Strawberry, Sainfoin, Buckwheat. Note in each case the features which show that the specimen is a fruit and not a seed.

5. Examine and draw capsules of various kinds to show their methods of dehiscence.

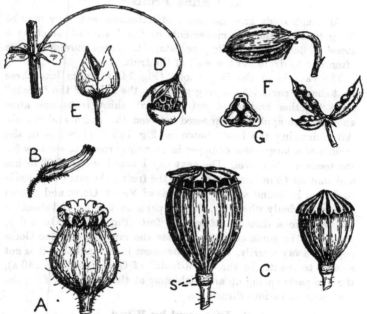

Fig. 142. CAPSULES of Campion (A), Mouse-ear Chickweed (B), old and young Poppy (C), Pimpernel (D), Figwort (E), Dog Violet (F). G, Violet fruit cut across. *s*, scars of stamens and petals. (Aug. × 1½.)

CHAPTER XXVII

FRUITS AND THE MIGRATIONS OF PLANTS—Continued

The Dispersal of the Seeds of Dry Fruits

1. "Sling-Fruits."

Although seeds have no power of self-movement, they may be flung to a distance by movements of the fruit-wall, much as a stone is flung from a sling or catapult. These movements are often due to drying of the wall of the fruit.

The capsule of the Dog Violet (Fig. 142 F) splits into three boat-shaped parts. As drying proceeds, the sides of the "boats" come together and shoot out the hard, shiny seeds one after another, like apple pips squeezed between the finger and thumb. After drawing the fruit shown in Fig. 142 I placed it in the middle of a large sheet of paper in an empty room, to see how far the seeds would travel. The next day I found one seed 22 inches and another 40 inches distant from the fruit ; the rest of the seeds could not be found at all. The pods of Vetch, Gorse and Broom dehisce suddenly when dry, the two parts curling up and shooting the seeds to a distance of several feet (Fig. 141 E). On a dry, sunny day in summer you can hear the ripe pods of the Gorse popping away merrily, but the movement is so rapid that it is not so easy to see it. In the "split-fruit" of Geranium (Fig. 140 A), the five parts spring up and, dehiscing at the same time, fling the five seeds in various directions.

2. Dispersal by Wind.

Dust-like seeds. Wind has little effect on very large seeds such as those of Horse Chestnut, but some seeds are so small that they can be blown about like dust, often for long distances, e.g. the seeds of Orchids, Heaths and Foxglove. Of course these dust-like seeds have the disadvantage that they contain very little reserve food, and when they germinate may find it difficult to compete with seedlings with plenty of food. On the other hand, some Orchids may produce hundreds of thousands of these minute seeds in a single fruit, and if only one or two succeed, that may be all that is required.

Censer-mechanisms. Ripe capsules and follicles often have stiff, erect stalks, on which the fruit is so placed that after dehiscence the seeds remain in a deep cup (Figs. 141 A, 142 A, B, C, E). The seeds are shed only when the fruit is shaken or jerked by the wind or a passing animal. As a rule a few seeds are shed at a time, and flung or else carried by the wind to some distance

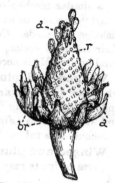

Fig. 143. CENSER-MECHANISM OF DAISY. *a*, achenes; *r*, receptacle; *br*, bracts. (Aug. × 3.)

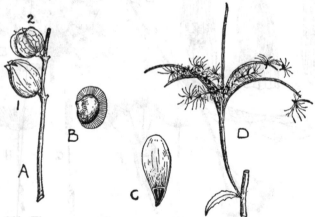

Fig. 144. WINGED AND PLUMED SEEDS. A, fruits of Yellow Rattle; 1, with, 2, without calyx. B, winged seed. C, winged seed of Spruce Fir. D, fruit and plumed seeds of Willow-herb. (Aug. A, D × 3/4; B × 2; C × 1.)

from the parent. The one-seeded fruits of the Daisy are scattered by a similar mechanism (Fig. 143). Such mechanisms are often called "censers," because of the swaying movements necessary to dislodge the seeds. Censer-mechanisms resemble sling-fruits in their mode of action, except that in the former the agency of wind or animals is necessary.

Winged and plumed seeds. Thin, flat wing-like edges which act like sails are found in some seeds, e.g. Yellow Rattle and Spruce Fir (Fig. 144 A—C). The seeds of Poplars, Willows and Willow-herbs (Figs. 113 H, 144 D) are plumed, i.e. there is a parachute-like tuft of hairs at one end of the seed.

Winged and plumed fruits. Like seeds, one-seeded fruits or parts of fruits may be either winged or plumed. The fruits of

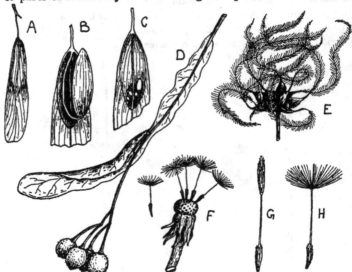

Fig. 145. WINGED AND PLUMED FRUITS. A, Ash. B, single seed of Ash. C, young Ash fruit with 4 ovules, only one of which will develop into a seed. D, Lime. E, Clematis. F—H. Dandelion. (C, June, the rest Sept. A, D, E, F × 3/4; B, C ×1; G, H ×1½.)

several of our trees, e.g. Ash (Fig. 145 A—C) and Elm (Fig. 100 F), have broad wings developed from the ovary wall. The Sycamore is similar, but has two wings (sometimes more), i.e. one to each one-seeded portion (Fig. 18). The Lime has a wing-like bract (Fig. 145 D), and is unusual in having several fruits dispersed by the aid of a single wing.

Wild Clematis or Old Man's Beard has a number of achenes developed from one flower (Fig. 145 E). As the fruits ripen, the styles grow into long feathery plumes, which assist in dispersing the fruits. The Dandelion has a tuft of hairs called the **pappus** (supposed to represent the calyx) at the top of each beaked achene (Fig. 145 F). You should try to follow for yourself the history of the flowers and fruits of the Dandelion. At first the young flowers in the bud are covered by the calyx-like involucre of bracts (Fig. 146 A). Then the flowers open (Fig. 125), and after pollination (p. 157) the corollas, stamens and styles fall off, the bracts closing over and sheltering the developing fruits; in this way we have a second bud-like stage (Fig. 146 B). Soon the bracts fold back and expose the ripe fruits (Fig. 145 F). At first the pappus is erect, and you can see that the hairs are seated on a minute green, watery swelling (Fig. 145 G). As this little swelling dries and disappears, the hairs are dragged downwards, and the parachute opens (Fig. 145 H); the fruits are now ready to be dispersed. Plumed fruits are also found in Groundsel and Thistles, and in Bulrush and Cotton Grass.

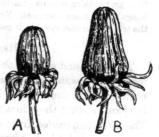

Fig. 146. DANDELION. A, inflorescence bud. B, a later stage with young fruits. (Aug. × 1.)

Except perhaps in censer-fruits, most of the seeds or fruits described above are small and light, and *their surface is very large compared with their weight.* This means that when once the seed or fruit is in the air, *it falls rather slowly to the ground,* which gives the wind more time to blow it away. You will see too that the usual position of flowers and fruits near the top of the plant (p. 164) aids not only pollination but also seed-dispersal. As a rule, winged fruits and seeds are not transported very far by wind. Plumed fruits and seeds can be carried to much greater distances, and probably dust-like seeds furthest of all. Winged fruits are indeed usually found in tall plants such as trees, where they have a long way to fall before reaching the ground. The wing causes the fruit to spin round and round as it falls, and so to fall more slowly. You should test the truth of this statement by the following experiment:

EXP. 33. Collect a number of winged fruits such as Sycamore, Ash or Elm, and cut off the wings from half of them. Then drop a few at a time from an upstairs window, starting equal numbers of fruits with and without wings at the same moment. You should ask a friend to stand underneath the window to observe which get to the ground first.

3. Dispersal by water.

The fruits and seeds of the Flowering Rush (Fig. 70), sedges and many other water-side plants are often dispersed by water currents. Many such fruits have air-chambers which make them buoyant, so that they float for a considerable time before sinking. The large one-seeded fruit of the tropical Coconut Palm, with its thick fibrous husk, is often carried for great distances by ocean currents. By this means the Coconut Palm has found its way to many islands in the South Seas.

4. Dispersal of the seeds of dry fruits by animals.

Burr-fruits. About 30 kinds of British plants have dry fruits provided with hooks which cling to the coats of passing animals, such as sheep, goats or dogs. In this way the seeds are often transported to a considerable distance. Several British "burr-fruits" are shown in Fig. 147. In Burdock (A, B) the bracts of the flowers are hooked, and in Agrimony (C) the hooks are on the floral receptacle. The Bur-Marigold (D) has a hooked pappus (calyx), while in Cleavers (E), Enchanter's Nightshade, and

Forget-me-not the hooks are all over the fruit coat. In Avens (Geum) the hooks are formed in a curious way. The flower is apocarpous, and each style has a kink in the middle. Later, as the fruit ripens, the style breaks at this kink, and a sharp hook is formed (Fig. 148).

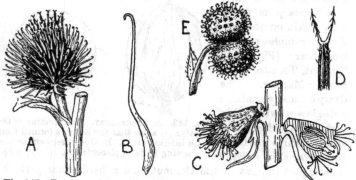

Fig. 147. Burr-fruits. A, Burdock. B, single bract of ditto. C, Agrimony (fruit on right shown in section). D, Bur-Marigold. E, Cleavers. (Aug. A × 3/4; the rest × 2.)

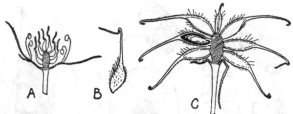

Fig. 148. Avens. A, longitudinal section through flower. B, a carpel. C, section showing fruits. Floral receptacle shaded. (Aug. A, C × 2; B × 4.)

Burr-fruits are common in many countries, and it is an interesting fact that they occur only on herbaceous plants with which hairy animals are likely to come in contact: they are never found on trees.

SUCCULENT FRUITS, AND THE DISPERSAL OF THEIR SEEDS

There are two common kinds of fleshy fruits, the berry and the drupe. The **berry** usually has several seeds (only one in the Date), each covered by a hard, protective testa. The pericarp is fleshy with a skin on the outside. Examples are the Gooseberry (Fig. 149), Currant, Tomato, Vegetable Marrow. The **drupe** usually contains a single seed with a thin testa. The pericarp consists of three parts; a thin outer skin, a fleshy

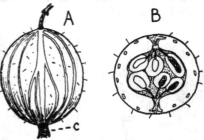

Fig. 149. A, GOOSEBERRY. The remains of the calyx (c) show that the berry is formed from an inferior ovary. B, Gooseberry cut across, showing parietal placentation. (July, × 1¼.)

middle part (fibrous in the Coconut) and a hard inner part, the "stone" or **endocarp,** which surrounds and protects the seed. In the Plum (Fig. 150 D), Cherry, Walnut and Coconut a single

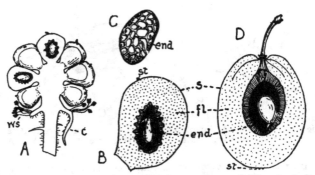

Fig. 150. Vertical sections through (A) BLACKBERRY, and (B) a single drupelet. C, Blackberry pip. D, vertical section through PLUM. *s.* skin; *fl,* fleshy layer; *end,* endocarp; *st,* remains of stigma; *c,* calyx; *ws,* withered stamens. (Aug. A × 2½; B, C × 7½; D × 3/4.)

drupe is formed from each fertilized flower; but in the Black-berry (Fig. 150 A—C) and Raspberry the flower is apocarpous, each ovary giving rise to a separate **drupelet.**

The seeds of succulent fruits are commonly scattered by animals, especially birds, which feed on the ripe fruits. Like the honey of flowers, the food in the fleshy pericarp is not used by the plant itself, but is a means of attracting animals which pollinate the flowers or disperse the seeds of the plant. Large fruits (e.g. Plum or Cherry) may be carried away by birds and eaten at leisure. The berries of the semi-parasitic Mistletoe (p. 102) are sticky, and the bird gets rid of the seeds by rubbing its beak against a con-venient bough. Many of the smaller berries and drupelets are, however, eaten bodily by birds. Some birds can grind up even the hardest seeds in their muscular gizzards; but with many fruit-eating birds, such as Blackbirds and Thrushes, the seeds, protected by the hard testa or endocarp, pass unharmed through the body of the bird. In this way, though succulent fruits containing several seeds are indehiscent, their seeds may be at the same time separated from one another and scattered.

False fruits. The succulent part of some fruits is formed partly or entirely from the floral receptacle. For instance, in the

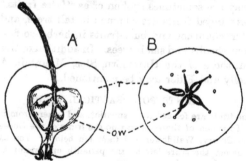

Fig. 151. YOUNG APPLE, cut vertically (A), and across (B). *ow*, ovary wall; *r*, fleshy receptacle. (Aug. ×1.)

Apple and Pear the horny "core" is the ovary wall, the soft pulpy portion being chiefly if not entirely receptacle (Fig. 151). The fleshy part of a Strawberry or a Rose hip (Fig. 152) is entirely receptacle and strictly speaking does not belong to the fruit at

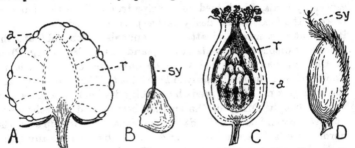

Fig. 152. STRAWBERRY (A) AND ROSE HIP (C) cut vertically. B, a Strawberry achene. D, ditto of Rose. *r*, fleshy receptacle; *a*, achenes; *sy*, style. (Aug. A, C ×1; B ×7; D ×4½.)

all, the true fruits in both cases being a collection of achenes. The achenes are enclosed by the receptacle in the Rose and outside it in the Strawberry, so that a Strawberry is rather like a Rose hip turned inside out. Fruits in which the receptacle becomes fleshy in this way are sometimes spoken of as "false fruits."

Just as winged fruits are common in tall trees, and other wind dispersal-mechanisms and burr-fruits in herbs, so fleshy fruits are common in shrubs and small trees. In addition to the fleshy fruits described, those of the Hawthorn, Sloe, Mountain Ash, Guelder Rose, Holly and Elder may be mentioned.

HINTS FOR PRACTICAL WORK.

1. Some fruits are formed from superior, and others from inferior ovaries (p. 150). To which of these classes do the fruits of the Gooseberry, Poppy, Sunflower, Cherry, Wallflower and Campion belong? You can tell this by finding whether the scars left by the perianth and stamens (or else their remains) are below or above the ovary.

2. As far as you can, examine the working of various dispersal-mechanisms. E.g. see how many jerks with the finger it takes to throw all the seeds

out of a "censer" of Campion or Poppy, and how far the seeds travel. You might also bring some ripe "sling-fruits" into a warm dry room, and see what happens to the seeds.

3. Wet some dry, open capsules of Campion or Cowslip, and see if the teeth close the capsule again. Do these capsules close naturally in wet weather?

CHAPTER XXVIII

THE RELATIONSHIPS OF FLOWERING PLANTS

Species. We see all around us vast numbers of individual plants, some of which so closely resemble one another that we say they belong to the same kind or **species.** For instance, all the Daisies in our fields belong to one species, and all the individual

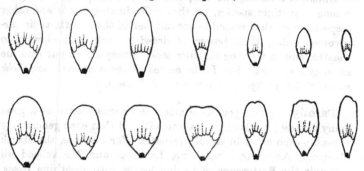

Fig. 153. VARIATION IN SHAPE AND SIZE OF PETALS OF LESSER CELANDINE.
The petals were selected from different flowers. (Apr. × 1.)

Lesser Celandines of our woods and hedgerows belong to another species. But with plants as with human beings, horses or dogs, however similar the individuals which make up a species may be, they are never exactly alike. Leaves and other parts may vary in shape, size and number, e.g. leaves of Dandelion (Fig. 39) and petals of Lesser Celandine (Fig. 153).

Genus. When several distinct species are constructed on the same general plan, and have a good many points in common, they are put into the same genus. For example, the Lesser Celandine, Creeping Buttercup, Meadow Buttercup, Bulbous Buttercup and Water Crowfoot have different leaves, roots, etc., but similar flowers and fruits. The flowers of all these plants are apocarpous and hypogynous; there is a nectary at the base of each petal (Fig. 108 *n*), and the fruits are achenes (Fig. 112). So these plants are put in different species but in the same genus. Naturalists have given all plants and animals a Latin name of two words, the first being the name of the genus, the second the name of the species. The Latin names of the five plants whose common names are given above are: *Ranunculus Ficaria*, *R. repens*, *R. acris*, *R. bulbosus* and *R. aquatilis*.

Amongst animals, the Cat, Lion, Tiger, Leopard and Jaguar belong to distinct species, for they differ in size, colour and other respects. But they resemble one another in their teeth, their carnivorous habit, their power of drawing back their claws into sheaths, and in walking on their toes, so they are all put into the same genus, *Felis*. The Latin names of these animals are *Felis catus*, *F. leo*, *F. tigris*, *F. pardus* and *F. onca*.

Family. When you have made the acquaintance of a good many different kinds of plants, you will find that even genera may resemble each other in certain respects. For instance, the Marsh Marigold, Anemone, Columbine, Larkspur and Monkshood all resemble the Buttercups in having leaves with sheathing bases, and apocarpous, hypogynous flowers with numerous stamens. None of these plants, however, have the nectary at the base of each petal which we find in the genus *Ranunculus*. Further, the Marsh Marigold and Anemone have only one set of perianth leaves, Columbine has five spurred petals, while Larkspur and Monkshood have irregular flowers. Because of these and other differences we put these plants into separate genera (*Caltha*, *Anemone*, *Aquilegia*, *Delphinium* and *Aconitum*). But because of their resemblances we put them, together with the genus *Ranunculus*, into one family, which we call the Ranunculaceae.

Dicotyledons and Monocotyledons. Nearly all foliage leaves belong to one or other of two types, (1) the net-veined type (Fig. 7), and (2) the long, narrow, grass-like type with parallel veins (Figs. 9, 10). The Ranunculaceae, Groundsel, Sycamore and many others belong to the first, and grasses, Daffodil, Bluebell and Crocus to the second type. We find further that plants with net-veined leaves usually have an embryo with two cotyledons, the parts of the flower often in fives, and a distinct calyx and corolla. On the other hand, plants with grass-like leaves generally have an embryo with one cotyledon, the parts of the flower in threes, and calyx and corolla like each other.

All ordinary flowering plants belong to one or other of these two groups, which are called respectively, from the number of their cotyledons, the **Dicotyledons** and **Monocotyledons.**

Both Dicotyledons and Monocotyledons include many families, and of course still more genera and species. You need not trouble much about the families of plants until you know a good many different species (see p. 193). But for purposes of reference a few families are mentioned below.

A. DICOTYLEDONS WITH FREE PETALS.

1. The Buttercup family (*Ranunculaceae*).

In this family the flowers are hypogynous, and all the floral leaves are free from one another. The stamens generally ripen before the stigmas. The nectaries are often modified petals (p. 162). The fruits are either achenes or follicles. Most members of the family are more or less poisonous.

2. The Wallflower family (*Cruciferae*).

The name Cruciferae (i.e. cross-bearers, Lat. *crux*, cross, *fero*, I bear) refers to the cross-like arrangement of the petals (Fig. 12). The flowers in this family are hypogynous, with a structure similar to that of the Dame's Violet (Chap. II). The fruit is a peculiar form of capsule; if long it is a **siliqua** (Fig. 14), if short a **silicula** (Fig. 154). When the fruit dehisces, the ovary wall breaks

away in two pieces from below upwards (Fig. 14). The seeds are either left behind and jerked away somewhat as in censer-fruits (Fig. 154), or else the fruit is a "sling-fruit."

To this family belong the Dame's Violet, Stock, Wallflower, Flixweed (p. 169), Cress, Radish; also the genus Brassica, which includes the Mustard, Cabbage, Cauliflower and Turnip. None of the Cruciferae are poisonous.

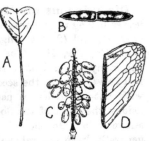

Fig. 154. SILICULA OF SHEPHERD'S PURSE. A, fruit. B, fruit cut across. C, seeds exposed after shedding of fruit wall, half of which is shown at D. (July, A ×1½; B—D ×3.)

3. The Bean and Pea family (*Leguminosae*).

The flowers are irregular and slightly perigynous (Fig. 156 A), in general resembling those of the Broad Bean (Fig. 15); the fruit is a legume (Fig. 141 D).

The flowers are pollinated by Bees. The Bee alights on the "wing" petals, and the weight of its body pushes them down, together with the "keel," which is interlocked with the wings

(cf. Fig. 155). This exposes stigma and anthers, both of which touch the under surface of the Bee. Self-pollination may occur in some cases; in this country the Sweet Pea is usually self-pollinated. The Broom has an explosive pollen flower, with 5 short and 5 long stamens (all united by their filaments) which, with the long style, are at first enclosed in the keel (Fig. 157 B). The weight of a Humble- or Hive-bee "explodes" the flower, when the short stamens touch the under surface,

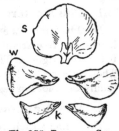

Fig. 155. PETALS OF SWEET PEA. *s*, standard; *w*, wings; *k*, the two keel petals. (Sept. ×1/2.)

and the stigma and long stamens the back of the insect. At first sight it would seem that the stigma could only be pollinated by pollen from long stamens. Probably, however, the short stamens

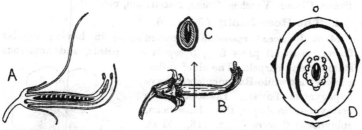

Fig. 156. SWEET PEA. A, longitudinal section of young flower. B, older flower with petals removed; the stigma projects further than in A. C, ovary and stamens cut across at arrow. D, floral diagram. (Sept. A, B ×1¾; C ×5.)

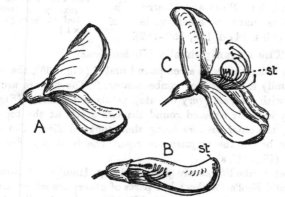

Fig. 157. FLOWERS OF BROOM. A, B, before, C, after "explosion." B, essential organs lying in keel. *st*, stigma. (July, ×1¼.)

give a second chance of pollination by a later visitor, for the style curls round till the stigma faces upwards, at the level of the short stamens (Fig. 157 c). The Broom flower only explodes once, but in Clover, Lupin, etc. the stamens and stigma retire into the keel again in readiness for another visitor.

The Leguminosae is a very large family, which includes Peas, Beans, Clover, Vetches, Gorse, Laburnum, etc.

4. The Rose family (*Rosaceae*).

Most Rosaceae resemble Ranunculaceae in having regular flowers with all parts free, 5 sepals and petals, and numerous stamens and carpels. The chief differ-

ence is that while Ranunculaceae have hypogynous flowers, Rosaceae have perigynous (in Apple, Pear, Hawthorn, epigynous) flowers (Figs. 118, 151 A). In Strawberry, Cinquefoil (Fig. 158), etc. there is an **epicalyx** of outer and often smaller leaves. There is only one carpel in Plum and Cherry. The Rosaceae have a great variety of fruits (Figs. 147 c, 148, 150–152).

Fig. 158. FLOWER OF CIN-QUEFOIL. *s*, sepal; *ep*, leaf of epicalyx. (Aug. × 1.)

5. The Parsnip family (*Umbelliferae*).

The inflorescence is a compound umbel (Fig. 135), the name of the family meaning the "umbel-bearers." The flowers are epigynous, with 5 sepals (very minute), petals and stamens (Fig. 119). Honey is freely exposed round the two styles at the top of the ovary, the chief visitors being short-tongued flies. The anthers wither before the stigmas are ripe. The fruit is a dry "split-fruit" (Fig. 140 B).

Most Umbelliferae are poisonous, e.g. Hemlock, Water Drop-wort and Fool's Parsley ; but parts of others are edible, e.g. roots of Parsnip and Carrot, leaves of Parsley and leaf-stalks of garden Celery

B. DICOTYLEDONS WITH JOINED PETALS.

6. The Primrose family (*Primulaceae*).

The flowers are regular and hypogynous, with 5 sepals, petals, stamens and carpels. The stamens are joined to and opposite the petals, and the ovary has free-central placentation (Fig. 123). Fruit a capsule dehiscing by teeth (by a lid in Pimpernel, Fig. 142 D). Examples are Primrose, Cowslip, Oxslip and Water Violet.

7. **The Dead Nettle family** (*Labiatae*).

The Labiatae (Lat. *labia*, a lip) have two-lipped, irregular, hypogynous flowers (Figs. 114, 126), 2 long and 2 short stamens and an ovary of 2 carpels which develops into a "split-fruit" of 4 nutlets (Fig. 140 D). The stem is square in cross section, and the leaves are in opposite pairs.

The leaves often contain scented oils, e.g. Lavender, Thyme, Rosemary, Marjoram, Sage and Mint.

8. **The Snapdragon family** (*Scrophulariaceae*).

The flowers often resemble those of Labiatae, but are not two-

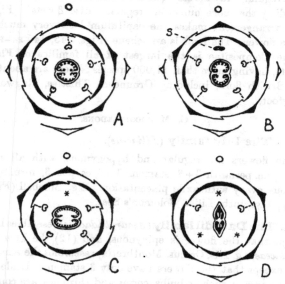

Fig. 159. FLORAL DIAGRAMS OF MULLEIN (A), FIGWORT (B), FOXGLOVE (C), SPEEDWELL (D). *s*, a barren stamen or "staminode" (cf. Fig. 127).

lipped in all cases; 4 stamens are usual (Figs. 128, 129, 159 B, C), but the Speedwell has 2 and Mullein 5 (Figs. 124, 159 A, D). The ovary has 2 carpels with axile placentation and many ovules (Fig. 159); fruit a capsule. The chief difference between this family and Labiatae is in the structure of the ovary. Most Scrophulariaceae are poisonous, e.g. Foxglove.

9. The Daisy and Dandelion family (*Compositae*).

The name Compositae (Lat. *compositus*, compound) has been given to this family because of the so-called "compound flowers," each of which is really an inflorescence (capitulum) of small flowers. The epigynous flowers are constructed on the same plan as the Dandelion (Fig. 125), but in many cases (e.g. Daisy) there are irregular "ray florets" (with no stamens) on the outside, surrounding the more numerous regular "disc florets" (Fig. 136). This arrangement makes the capitulum look very much like a single flower. The fruit is an achene (Figs. 143, 145 F—H).

The Compositae is the largest of all families of Flowering Plants, having more than 11,000 species; there are about 109 in the British flora, including Groundsel, Thistles, Ragworts and Coltsfoot.

C. MONOCOTYLEDONS.

10. The Lily family (*Liliaceae*).

The flowers are regular and hypogynous, with all parts in threes, i.e. perianth $3 + 3$, stamens $3 + 3$, carpels 3, forming a syncarpous ovary with axile placentation. E.g. Bluebell (Fig. 117), Tulip, Hyacinth, Lilies, Solomon's Seal.

11. The Daffodil family (*Amaryllidaceae*) resembles Liliaceae except that the flower is epigynous, and (12) the Iris family (*Iridaceae*, e.g. Iris, Crocus, Montbretia) resembles the Amaryllidaceae except that the flowers have only 3 stamens. Underground storage organs, such as bulbs, corms and rhizomes, are common in these three families (Figs. 84, 86, 87, 101 A).

The Use of a "Flora."

The various species of wild plants inhabiting any country make up what is known as the **flora** (from Lat. *flos*, a flower) of that country. The word flora has also come to be used for a book containing a list and descriptions of the plants found in a country or other geographical area. A number of such "British Floras" have been written, perhaps the simplest of which for a beginner is *A School Flora*, by W. M. Watts (Longmans, Green and Co.). Instructions are given in the preface to this book how to use the "keys" by means of which, after some practice, you can find out the name of any common wild flowering plant, and a good many of the rarer ones. This is very interesting work, for getting to know new plants is rather like making new friends. Of course you should not be content merely to learn the name of a plant, but should try to find out as much as you can about it, where it grows, the form of its leaves, the structure of its flowers and fruits, and so on. Gradually you will become familiar with many of our commoner species, and then you can group them up into their families. You will find that in some families, such as the Cruciferae, Leguminosae, Umbelliferae, Labiatae and Compositae, the structure of the inflorescence, flower and fruit is more or less similar throughout a whole family, so that the family is easy to recognize. Other families show a much wider range of structure. The Ranunculaceae, for instance, includes flowers of very varied form as regards the perianth (p. 186), though in each case the essential organs of the flower are constructed on the same plan. The Rosaceae and the Scrophulariaceae too are sometimes a little difficult to recognize, on account of the wide range of structure in their flowers and fruits (pp. 190, 191–2). But any little difficulty of this sort can be overcome by perseverance and practice.

HINTS FOR PRACTICAL WORK.

1. Collect specimens of common wild flowers when you go for walks, and get to know as many kinds as you can. It is a good plan when you find a plant you do not know, to make a short list of these features which you think may help you to recognize the plant again.

2. In order to understand what is meant by a species, collect two or three kinds of Buttercup, or Vetch or Speedwell, and examine them very carefully. Make as full lists as you can of the characters in which the flowers, fruits, leaves and other parts agree and differ.

APPENDIX

THE NAMES OF PLANTS

Most of the plant-names in this book are the English names in common use in this country. Sometimes a plant has several names, used perhaps in different parts of the country. The following, for instance, are some of the names which have been given to the wild Arum : Lords and Ladies, Cuckoo-pint, Wake Robin, Wakepintle, Aaron, Calves-foot, Starchwort. Other countries too have their own plant-names, so that a plant which grows in many countries may be called by a great number of different names. Many years ago, therefore, naturalists thought it would be a good thing to give each species of plant and animal one scientific name, by which it would be known to the naturalists of all countries. At that time it was the custom for scientific men to write their books in Latin, so the scientific names given to plants and animals were either Latin (or sometimes Greek) words, or words put into a Latin form (cf. p. 186). Scientific books are no longer written in the Latin language, but the custom of using Latin names has been found so useful that it still persists.

The scientific names of a number of the plants mentioned in this book are given below. The list may be useful to those who wish to know the Latin as well as the English names. You will see from some of the derivations given, that the name often refers to some peculiarity of the plant or of its habitat. The same is true of many English names, e.g. Buttercup, Bluebell, Sundew, Meadow Sweet.

An asterisk (*) after the name of a plant means that the plant has been introduced into this country from abroad ; all the others are native British plants.

English Name.	Latin Name.
Alder	*Alnus glutinosus.*
Anemone (Wood)	*Anemone nemorosa* (Lat. *nemus*, a wood or grove).
Arrowhead	*Sagittaria sagittifolia* (Lat. *sagitta*, an arrow; *folium*, a leaf).
Ash	*Fraxinus excelsior.*
Bean (Broad)	*Vicia Faba** (Lat. *vicia*, a vetch).
Bean (French or Kidney)	*Phaseolus vulgaris** (Lat. *vulgaris*, common).
Beech	*Fagus sylvatica* (Lat. *silva* or *sylva*, a forest).
Birch	*Betula alba.*
Bladderwort (Common)	*Utricularia vulgaris* (Lat. *utriculus*, a small skin or bladder).
Bluebell	*Scilla nutans* (Lat. *nutans*, nodding).
Bramble or Blackberry	*Rubus fruticosus* (Lat. *rubus*, a bramble-bush).
Broom	*Cytisus scoparius* (Lat. *scopae*, a broom).
Buttercup (Creeping)	*Ranunculus repens* (Lat. *repens*, creeping).
Buttercup (Meadow)	*Ranunculus acris.*
Butterwort	*Pinguicula vulgaris.*
Candytuft	*Iberis amara**.
Celandine (Lesser)	*Ranunculus Ficaria.*
Chickweed	*Stellaria media* (Lat. *stella*, a star).
Cleavers or Goosegrass	*Galium Aparine.*
Clover (White)	*Trifolium repens.*
Cock's-foot Grass	*Dactylis glomerata.*
Columbine (wild)	*Aquilegia vulgaris.*
Daffodil (wild)	*Narcissus Pseudo-narcissus.*
Daisy	*Bellis perennis.*
Dame's Violet	*Hesperis matronalis**.
Dandelion	*Taraxacum Dens-leonis* or *officinale* (Lat. *dens*, a tooth; *leo*, a lion).
Elm	*Ulmus campestris* (Lat. *ulmus*, an Elm-tree; *campestris*, dwelling in plains).
Enchanter's Nightshade	*Circaea lutetiana* (from *Circe*, the enchantress).

Eyebright	*Euphrasia officinalis* (Lat. *officinalis*, of the shops; applied to medicinal plants sold in shops).
Figwort	*Scrophularia nodosa*.
Flixweed	*Sisymbrium Sophia*.
Flowering Rush	*Butomus umbellatus*.
Foxglove	*Digitalis purpurea* (Lat. *digitus*, a finger, *purpureus*, purple).
Gooseberry...........................	*Ribes Grossularia*.
Gorse, Furze or Whin	*Ulex europaeus*.
Gout Weed or Bishop Weed ...	*Aegopodium Podagraria* (Lat. *podagra*, the gout).
Groundsel	*Senecio vulgaris*.
Harebell	*Campanula rotundifolia* (Lat. *rotundus*, round; *folium*, a leaf).
Hawthorn or May..................	*Crataegus Oxyacantha* (Gk. *oxys*, sharp; *acantha*, a thorn).
Hazel	*Corylus Avellana* (Lat. *corylus*, a Hazel or Filbert tree).
Hogweed or Cow Parsnip	*Heracleum Sphondylium*.
Horse Chestnut.....................	*Aesculus Hippocastanum**.
Larkspur	*Delphinium** (several species are cultivated).
Lilac	*Syringa vulgaris**.
Lily of the Valley..................	*Convallaria majalis*.
Lime	*Tilia europaea**.
Lords and Ladies	*Arum maculatum* (Lat. *macula*, a spot).
Maize or Indian Corn	*Zea Mays**.
Meadow Sweet	*Spiraea Ulmaria*.
Monkshood or Aconite............	*Aconitum Napellus*.
Nasturtium (Garden)	*Tropaeolum** (the climbing one is *T. majus*).
Oak	*Quercus Robur* (Lat. *quercus*, an Oak-tree; *robur*, hardness, strength, also used for Oak).
Pea (Garden)	*Pisum sativum** (Lat. *pisum*, a Pea; *sativus*, that which is sown).
Pea (Meadow)	*Lathyrus pratensis* (Lat. *pratensis*, found in meadows).

Pea (Sweet) .... *Lathyrus odoratus** (Lat. *odoratus*, fragrant).
Plantain (Great) .... *Plantago major.*
Plantain (Ribwort) .... *Plantago lanceolata.*
Pondweed (Shining, Fig. 63) ... *Potamogeton lucens* (Gk. *potamos*, a river; Lat. *lucens*, bright, shining).

Privet .... *Ligustrum vulgare* (Lat. *ligustrum*, Privet).

Sensitive Plant .... *Mimosa pudica** (Lat. *pudicus*, modest).
Shoreweed .... *Littorella lacustris* (Lat. *littus*, a shore; *lacus*, a lake).

Snapdragon .... *Antirrhinum majus*.*
Solomon's Seal .... *Polygonatum multiflorum* (Lat. *multus*, many; *flos*, a flower).

Speedwell (Germander) .... *Veronica Chamaedrys.*
Strawberry .... *Fragaria vesca* (Lat. *fragum*, a Strawberry).
Sundew (Round-leaved) .... *Drosera rotundifolia* (Gk. *droseros*, dewy).
Sycamore .... *Acer Pseudo-platanus*.*

Violet (Dog) .... *Viola canina* (Lat. *canis*, a dog).
Violet (Marsh) .... *Viola palustris* (Lat. *palus*, a marsh).
Virginian Creeper .... *Ampelopsis Veitchii** (Gk. *ampelos*, a Vine).

Water Crowfoot..................... *Ranunculus aquatilis* (Lat. *aquatilis*, found in water).
Water Lobelia .... *Lobelia Dortmanna.*
Water Milfoil....................... *Myriophyllum spicatum* (Gk. *myrioi*, ten thousand; *phyllon*, a leaf).

Wood Sage........................... *Teucrium Scorodonia.*

Yellow Rattle....................... *Rhinanthus Crista-galli* (Lat. *crista*, a comb; *gallus*, a cock).

INDEX

Printed in the United States
By Bookmasters